Molten Silicates
and their Properties

Molten Silicates
and their Properties

by B. Lőcsei

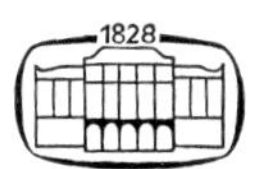

Akadémiai Kiadó, Budapest 1970

This book is the enlarged English version
of the original Hungarian edition entitled
"Olvasztott szilikátok és tulajdonságaik"
published by Akadémiai Kiadó, Budapest

Translated by
F. Sós

Contents

Chapter I

Introduction

The rapid technical development in the 20th century has raised demands for the preparation and application of more and more new structural materials. Of the non-metallic materials, besides the various plastics, the importance of the application of glass and of other silicate products is on the increase. In the latter group belong the substances which may be classified under the collective term "molten silicates" and which up to the first half of the 1950's were exclusively the by-products of rock processing. Earlier the crystallization of glass was an undesirable source of defects in glass manufacture. For this reason the study of crystallization ability is of special importance. At the same time the efforts to avoid crystallization threw light on crystallization conditions both from the theoretical and practical aspect. The information gained in this field may be utilized for both reducing and promoting crystallization tendencies whereby methods to control crystallization ability are being evolved.

RÉAUMUR was the first to attempt a transformation of glass into a crystalline substance, but all his experiments were frustrated by the deformation of the material during the crystallization process.

In the last 20 years RÉAUMUR's old idea was brought up again, but now enriched by significant theoretical silicate chemical content to produce ceramic products by way of glass crystallization. Under the collective denomination of "molten silicates" we now understand ceramic materials produced from one or more base materials by melting. The material is molded in its liquid state like any other viscous material and converted in its bulk into a crystalline state by thermal treatment during cooling to avoid any significant change in shape, that is deformation, during thermal treatment [1].

Molten silicates may be classified into two groups:

(a) Crystallization of the first type begins at a higher temperature during molding. The thermal treatment after molding only completes this process. This type of crystallization may be considered primary crystallization from the aspect of the "thermal past" of the substance. Crystallization begins during cooling when the crystallization temperature range is primarily reached, after having reached the liquidus temperature, that is the material has cooled below the liquidus temperature.

Into this group belong mainly molten silicate products obtained by melting of rocks and slags. This method is applied primarily to the processing of basalt and diabase [2–14].

(b) In the second group belong the molten silicates called by HINZ [15] vitroceramics. These are characterized by being shaped in the glassy state and by the fact that their crystallization begins during thermal treatment following forming at a lower temperature and proceeding towards higher temperatures where the crystallization process is terminated. This type of crystallization operation may be considered secondary, since during the crystallization process the substance reaches a second time the temperature range of crystallization. In addition, nucleators are usually added in the course of preparation. Nucleators are small quantities of additives which bring about deformation-free crystallization. This, that is nucleation kinetics, is the theoretical surplus of silicate chemistry by means of which RÉAUMUR's idea could be realized. Nucleators at the same time provide a possibility of varying the composition between wider limits and of arriving at more favorable material structures.

In this book the author could of course not attempt to discuss all the problems of vitroceramics, firstly because of only a limited available space. His analysis will refer mainly to vitroceramics in the SiO_2–Al_2O_3–CaO–MgO–Na_2O system and to the mechanism of metal sulphide nucleation. He will also report his own results with heavy metal sulphide nucleating agents and his "micro-eutectic" principle. In the discussion of general theoretical problems he will also deal with certain aspects of TiO_2 nucleation and this mainly for the sake of comparison, e.g. in the discussion of nucleation, that is of heterogeneous crystallization mechanism, he will demonstrate the control of TiO_2 nucleation by means of sulphide, selenide and telluride, respectively.

Molten silicates owe their utilization as structural materials partly to their favorable mechanical and chemical properties. In addition because of some other advantages, e.g. dielectric properties, they play an important role in many industries. The basic principles of preparation of vitroceramics were evolved in the past 20 years and the commercial manufacture of various material types was commenced in the near past.

Chapter II

Historical

It appears from the available literature that basalt melting was first performed in 1777 in France by d'ARCEL who, in the course of studying the crystallization conditions of magma, melted basalt in a porcelain crucible [3]. At the end of the 18th century the geologist GOLLOW [2] carried out also some work on the melting of eruptive rocks, and found that depending on the conditions of cooling, molten basalt will solidify to form either a vitreous or a crystalline solid. Somewhat later HALL and WATT melted basalt in a crucible and in a small rotation furnace. FOUQUÉ and MICHEL-LÉVY [1, 3] also carried out some important work in this field. They prepared synthetically a product corresponding petrographically with basalt, and thus, it was demonstrated that the method of cooling has a role in the textural structure of eruptive rocks – besides the chemical composition. They stated that "the origin of eruptive rocks is the consequence of their melting, followed by a slow cooling". Their method at the same time raises the probability of producing such synthetic materials with compositions differing from that of natural basalt and their composition number adjusted so as to give the optimum application properties to the material. FOUQUÉ and MICHEL-LÉVY did not measure temperature in their experiments and their explanations concerning the phenomena are not right because of the basic concepts of silicate chemistry at that time were not yet clarified. Obviously they did not know the laws governing the solidification of heterogeneous systems [16].

In 1837, a German scientist, BISCHOFF performed experiments on basalt melting. He poured about 350 kg of basalt melt into a clay mold and succeeded in crystallizing the melt. ADCOCK was the first taking out a patent for invention on the melting of basalt and the processing of the melt in the middle of the past century. He wanted to make glass from molten basalt, but did not envisage the difficulty that the basalt melted by blowing may not be molded because of its crystallization on heating. Thus if the ball on the blowing iron is again immersed in the melt, the already blown material will crystallize as shown by Hungarian experiments, too. In the production of large articles cooling rate was not high enough to allow the material to cool in a homogeneous vitreous state and the formation of crystal nuclei made the material brittle. This explains why was ADCOCK's invention a failure and completely forgotten [3].

The Mauritian French doctor RIBBE carried out some melting experiments with French basalts, and established the conditions for their melting and thermal treatment. Later the engineer DRIN investigated the application possibilities of the material in industry. As a result of their work the Compagnie Générale de Bazalte was founded in Paris having two factories, one in the vicinity of Paris working with gas-heated furnaces and a daily capacity of 10 tons, the other in Auverigne with a capacity of 8 tons per day, and using electric melting furnaces [1, 6, 9].

In 1922 a basalt melting plant was established in Kahlenborn in Germany on the basis of French literature data and experience. In the 1920's the Soviet Union too began research on the melting of basalt, in 1926 GINZBURG in the Laboratory of the Rock Research Institute and FLORENSKII in the State Research Institute for Electrotechnique were working on the problem [2]. The laboratory experiments were soon followed by pilot plant trials on whose experience a factory with a yearly capacity of 5000 tons was built. The first diabase works started operation in Moscow in 1932. Experiments have shown the diabases from the Oniega district to have the most favorable technological properties. In Leningrad the Armenian andesite-basalt deposits were investigated and an appropriate technology for their processing worked out.

After World War II Czechoslovak specialists also began to be interested in molten silicates, since they had some earlier favorable experiences with molten basalt of German origin. They began their experiments in 1948 and within a few years had two plants operating with daily 16 and 10 ton capacities, respectively and today it has become necessary to plan for expansion. Poland set up a similar factory in 1954 [17–19].

In Hungary investigation of the production conditions of molten silicates began in the Glass Department of the Research Institute for Heavy Chemical Industries in 1951. Beside basalt melting experiments another type was also elaborated first on furnace slag base and later, after having clarified the role and essence of nucleation, a general process based on oxide synthesis and independent of the base material was envisaged. This work may be considered the first step in the production of vitroceramics or of ceramics produced by the nucleated crystallization of glasses. LŐCSEI and his research group have applied first in 1951, then in 1953 and 1957 for patents covering the production of secondarily crystallizing vitroceramics [20–23].

BECKER has prepared as far back as the 1930's primarily crystallizing molten silicates with high fluorine content [30, 31]. Later in 1952 WAGNER also produced crystallizing materials with high fluorine content.

STOOKEY applied in 1953 for patents covering pyroceramic and photoceramic materials which resulted from the development of photosensitive glasses. The material was prepared on lithium silicate, lithium magnesium aluminum silicate base with metal (Ag, Au, Cu) and TiO_2 nucleation [25–29]. In the Corning Works thorough research is in progress under STOOKEY's direction leading to four main types [28]. POLINSZKY and LŐCSEI have reported on the production possibilities of vitroceramics on furnace slag base [81]. In 1956 LUNGU [32, 33] published his results on a vitroceramic in the SiO_2–Al_2O_3–MgO–Na_2O system produced by fluoride nucleation. This was followed by types worked out in Great Britain as a result of the work of CLAYPOOLE, MACMILLAN, PARTRIDGE and HODGSON in the years between 1959 and 1963 [34–39].

In the Soviet Union a new type of vitroceramic material was marketed n 1960 under the trade-name "Sital" [40]. RINDONE studied the nucleation mechanism of lithium aluminum silicate crystallization when metallic platinum is added [41].

Since 1960 research is in progress in many institutes of many countries aiming at the preparation of new vitroceramics and at the clarification of silicate chemical processes [42–68]. In 1960 HINZ and WISHMANN produced a molten silicate of microcrystalline structure from copper slag glass obtained by melting and by the addition of Cr_2O_3.

HINZ and KNUTH studied the effect of small quantities of diverse additives on the properties of pyroceramic type materials in the Li_2O– Al_2O_3– SiO_2 system [69].

In the Institute in Hradetz-Kralove VOLDAN is engaged in the investigation of the silicate chemical principles of vitroceramic materials and BROUKAL in that of pyroceramic soldering vitroceramics [70, 71].

In the Soviet Union vitroceramics on furnace slag base are being produced since 1962, in Poland since 1964, in both countries production is based on principles which are in agreement with LŐCSEI's experimental results [72, 73]. VOGEL and GERTH [74] and later MAURER [75] clarified many details of the nucleation mechanism and the importance of homogeneous and heterogeneous catalysis.

The study of vitroceramics received particularly great impetus in the past years both from the theoretical and the engineering aspect producing a continuous succession of new results. Interest is focused on the kinetical and thermodynamical study of nucleation [173–179]. The properties of lithium aluminum magnesium silicates nucleated with phosphates were studied by PARTRIDGE and MACMILLAN who found a new method for the processing of vitroceramics [194, 196]. The development of vitroceramics which may be used as solders is to a major part also the merit of these authors [197]. HODGSON and MACMILLAN reported on neutron absorbent vitroceramics [198]. Control of crystallization tendency and within this of crystallization rate by means of the combined application of nucleating agents is again a theoretical result which will obviously bring about an extension of the composition of vitroceramics [171].

An interesting development of the research on vitroceramics is the improved strength by surface treatment. When vitroceramics immersed in some liquid salt are subjected to heat treatment, the surface layer will shrink as a result of ion exchange and a compressive stress will develop resulting in a considerable improvement of crushing strength. KARSTETTER et al. in the Corning Works have been engaged in the study of this problem [186–193].

Another recent result is the influencing of the crystallization mechanism to form on the surface of the vitroceramic material a crystalline layer of different structure in a single operation during crystallizing thermal treatment. This is another recent solution to improve the flexural strength of vitroceramics [171].

The new possibility of coloring vitroceramics has also extended the application possibilities of these materials [181, 184]. Materials with high thermal expansion coefficients in the SiO_2–P_2O_5–Al_2O_3–CaO–MgO system or vitroceramics with ZnO or BaO content can be coated with metals [180,

189]. Application to other ceramics and to metals has also opened new perspectives to these materials.

The importance of vitroceramics is indicated by the conferences on this subject such as were held in Toronto in 1961, or in Prague in 1967 in addition to a large number of national symposia. At the International Conference in London in 1968 a special section discussed the problems related to vitroceramics [181, 182, 183, 185, 195, 199]. As another sign of the importance of vitroceramics the Department of Glass Technology of the Mendeleev Institute in Moscow has changed its name to Department of Glass and Sital (vitroceramics) Technology.

Today, though vitroceramics are manufactured on the commercial scale, there are still many unsolved problems to interest the scientist, including the full possibilities of nucleation, the mechanism and kinetics of nucleation and heterogeneous catalysis, the effect of the composition of the base glass, refining of the structure, improvement of the physical and chemical properties of the materials and the revealing and extension of application possibilities.

Chapter III

Basic principles of molten silicate production

The production of the first group of molten silicates, that is of molten rocks, is based on the choice of a rock composition which after melting will have a crystallization ability to initiate the crystallization process in the course of molding. The low network forming and modifying cation ratio of the basalt and slag melts is in fact the primary cause of the high crystallization ability, while primary crystallization during cooling may be attributed to the presence of residual crystal nuclei in the melt [70]. This crystallization commencing at a high temperature will ensure later, when crystallization under the effect of heat treatment has been completed, the permanent molded shape of the material and its cooling without deformation or cracking.

In the case of vitroceramics on the other hand the composition of the base material has to be chosen to prevent crystallization of the melt during shaping. In this case the melt has to be solidified without crystallization. After molding a heat treatment according to a definite program will initiate the separation of the nucleating agent. This primary crystal phase, formed in the material at a lower temperature, will ensure in the course of subsequent heat treatment a deformation-free recrystallization.

Thus in both types the crystallization ability of the base material melt plays an important role to which the presence of nucleating agents contributes considerably, especially in the case of vitroceramics. The fundamental phenomena are essentially the same in both cases. The crystal residues in the rock melts – olivine, magnetite – act in fact like heterogeneous nucleating agents. The difference between the structures of the two types of materials may be interpreted from the relative values of the temperature gradient of the crystallization process and of the two parameters of crystallization tendency, namely of crystallization rate and the number of crystal nuclei. This explains that the texture of molten rocks is an agglomeration of larger single crystals that are found in vitroceramics. This appears quite clearly from the comparison of Figs 1 and 2 representing thin sections of a molten Kahlenborn basalt and a vitroceramic KM2 prepared by Lőcsei [1] from furnace slag. The magnifications are the same (500 ×).

The difference is even more striking when two thin sections of the materials are compared under 1500× magnification (Figs 3 and 4). On the slide of the molten basalt diopside twin crystals crystallized by leaf-like morphology are visible. One division in the figure is 1 μm. On the slide of the material prepared from furnace slag and called "crystalline synthetic stone" (KM2) the texture of far smaller structural elements is interrupted here and there by some larger crystals [1].

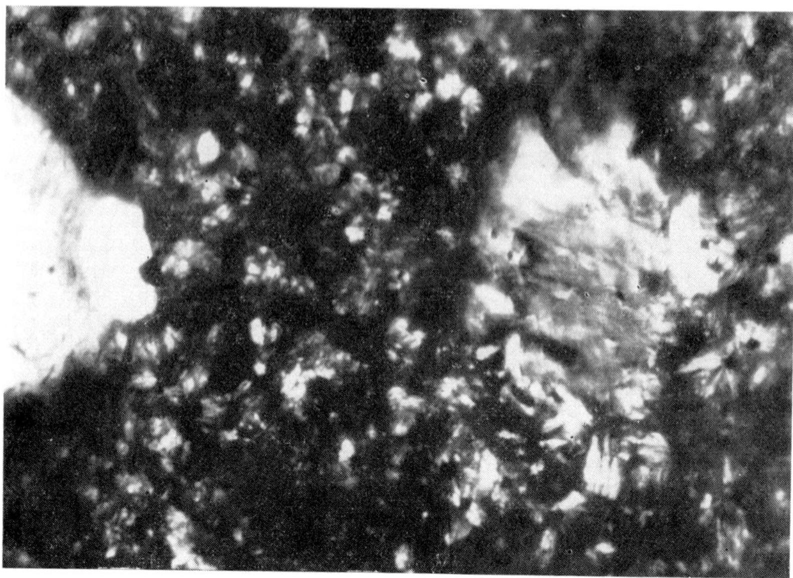

Fig. 1. Structure of molten Kahlenborn basalt. Microscopic picture of a thin section in transmitted polarized light. Magnification: $400\times$

Fig. 2. Structure of a vitroceramic material on furnace slag base. Microscopic picture of a thin section in transmitted polarized light. Magnification: $400\times$

FIG. 3. Structure of molten Kahlenborn basalt. Picture of the same slide as in Fig. 1 in polarized light. Magnification: 1500×, 1 division: 1 μm

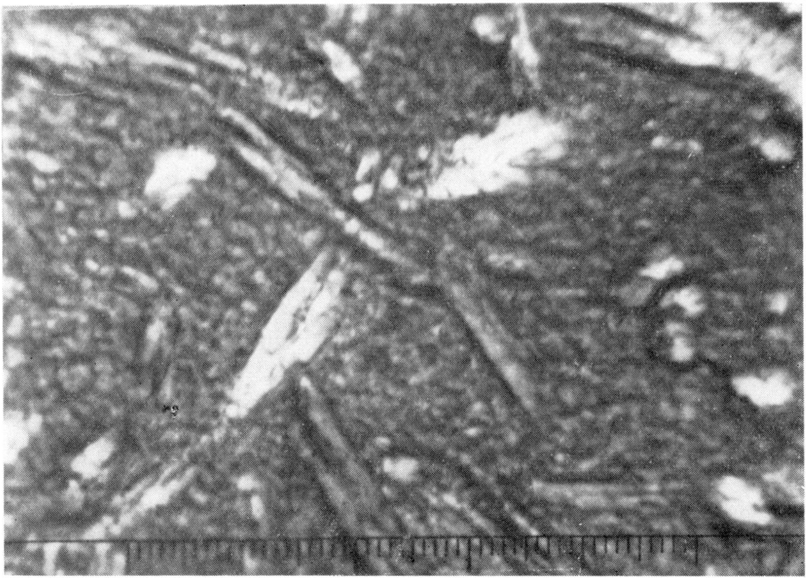

FIG. 4. Structure of a vitroceramic material on furnace slag base. Picture of the same thin section as in Fig. 2 in polarized light. Magnification: 1500×, 1 division: 1 μm

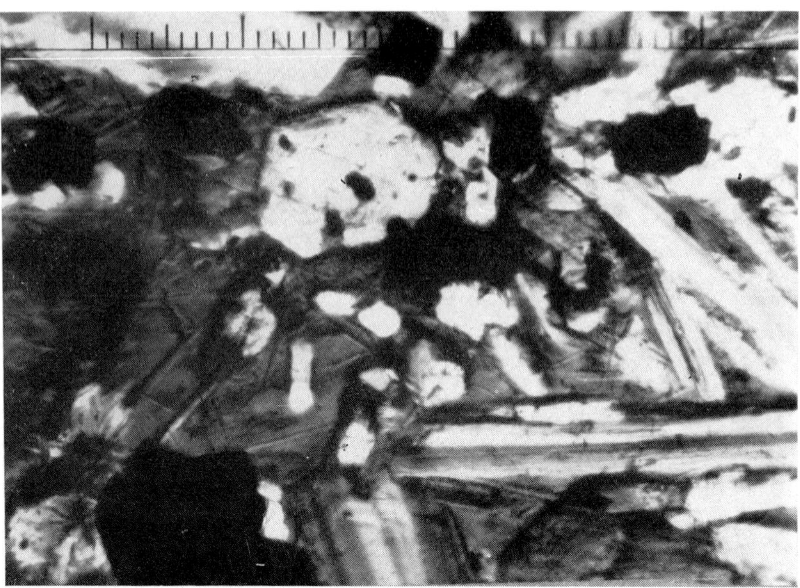

FIG. 5. Structure of Badacsony basalt. Microscopic picture of a thin section. Magnification: $400\times$, 1 division: 5 μm

The texture of vitroceramics consists of birefringent individual crystals of 0.5–3 μm dimension, the larger crystals are often sloping rod or double-rod shaped. In addition black opaque crystals of 0.5 μm diameter also appear in the texture. X-ray diffraction tests identified these latter as iron and manganese sulphide crystals, while chromium oxide crystals are also present. The larger crystals in Figs 2 and 4 have a hexagonal cross-section perpendicularly to their length, their extinction is sloping with different angles in different orientations. In this small dimension it is impossible to identify optically the minerals. Figure 4 is not a sharp picture, since the shape of the crystals is not regular. The dissolved large negative crystals with double optical axes and without preexisted faces are feldspars (Fig. 1). In Fig. 3 the leaf-like crystals form radial bunches according to the crystal form of magnetite which supports the earlier statements on crystal fragments. This characteristic crystal growth also confirms the role of magnetite as a heterogeneous nucleating agent. The development of the leaf-shaped crystals was made possible by the presence of an amorphous phase during crystallization. Well-developed crystal forms can separate from the melt only. The textures in Figs 1–4 were photographed in polarized light. Figure 5 shows for the sake of comparison a thin section of Badacsony basalt.

Beyond the circumstances of crystallization the material properties of molten silicates are also functions of the choice of the base material. The properties of vitroceramics may be varied between wide limits by the modification of their composition.

In the processing of molten silicates the following aspects must be allowed:

1. Cost of the base material – composition and choice of the raw materials
2. fusibility
3. moldability
4. choice of nucleating agents, the quality of nucleation
5. quantity of nucleating agents, combination ratio
6. possible control of crystallization rate
7. changes in specific gravity and volume, shrinkage during crystallization
8. mechanical strength
9. chemical resistance
10. changes in thermal expansion coefficient during crystallizing heat treatment up to the end state
11. heat shock resistance
12. proportion of the amorphous to the crystalline phase during crystallization
13. surface quality as a function of the nucleating agent and base glass composition
14. dielectrical properties
15. abrasion resistance

(A) Processing of petrous materials

The possibility of processing a rock or slag to obtain molten silicate, that is, a petrous product, is determined by the following conditions:

1. The rock must be homogeneous; that is, no fluctuations in its composition are permissible. It must contain no foreign matter, veins or occlusions of different compositions.
2. The rock must be pure, it must contain no wear products. It should be easy to melt, when a uniform fine structure is an advantage and coarse occlusion a disadvantage.
3. The viscosity of the melt should be low at the processing temperature, so that the melt should be easy to homogenize, to clarify and to cast.
4. Castings from the melt should crystallize well; the finer the texture the more preferable it will be.
5. Cast and crystallized products should have no tendency to crack.

The eruptive rocks may only meet the above requirements. According to their chemical composition they may be classified into the following groups:

(a) acidic eruptive rocks with a silica content above 65%,
(b) intermediary eruptive rocks with a silica content between 52 and 65%,
(c) basic eruptive rocks with a silica content less than 52%.

The higher the silica content the more quartz and feldspar is in the rock; the first has a detrimental effect on fusibility, the second raises the viscosity of the melt.

The above requirements are met by the basic eruptive rocks alone which contain no quartz and whose feldspar content is replaced to a considerable degree by easily melting pyroxenes. Because of their low silica content their viscosities at high temperatures are low and hence they are easy to cast, the high crystallization tendency is a result of low viscosity and high basicity. Of the eruptive rocks the effusive basic rocks are those primarily suited to the production of molten silicates; these rocks are the basalts and diabases.

TABLE 1/a

Average composition of various basalt types in weight per cent

Oxide	Melilite basalt	Melilitite	Nepheline basalt	Nephelinite	Limburgite	Augite	Nepheline basanite	Leucite basanite
	No. of analysis							
	5	5	26	9	14	6	16	10
SiO_2	35.72	37.56	39.87	41.17	41.25	42.25	44.20	45.55
TiO_2	4.78	2.66	1.50	1.35	1.59	2.52	1.64	2.33
Al_2O_3	6.56	10.08	13.58	16.83	12.03	16.26	15.64	14.97
Fe_2O_3	5.41	6.82	6.71	7.61	5.65	8.43	4.35	4.77
FeO	9.55	5.94	6.43	6.64	7.29	5.46	6.14	6.64
MnO	—	0.06	0.21	0.16	0.54	—	0.19	0.61
MgO	15.46	15.32	10.46	3.72	11.22	5.49	0.89	6.41
CaO	14.20	13.82	12.36	10.12	11.88	9.75	9.74	10.16
Na_2O	3.35	3.11	3.85	6.45	3.40	4.45	4.03	2.76
K_2O	1.67	1.53	1.87	2.49	1.30	1.92	1.83	4.04
H_2O	2.67	2.52	2.22	2.42	3.20	2.43	2.67	1.61
P_2O_5	0.63	0.58	0.94	1.04	0.65	1.04	0.68	0.15
(C)*	1.80	1.90	1.71	1.60	1.57	1.49	1.46	1.34

TABLE 1/b

Oxide	Basanites	Leucite basalt	Leucitite	Nepheline tephrite	Leucite tephrite	Tephrites	Basalts
	No. of analysis						
	26	12	9	4	20	24	161
SiO_2	44.64	46.18	46.90	46.91	49.90	49.14	48.78
TiO_2	1.95	2.13	1.22	1.81	0.16	1.00	1.39
Al_2O_3	15.35	12.74	16.33	15.25	16.94	16.57	15.85
Fe_2O_3	4.50	5.27	4.22	7.70	3.02	3.65	5.37
FeO	6.35	5.06	4.14	4.06	7.15	6.68	6.34
MnO	0.46	0.19	0.11	1.43	0.23	0.30	0.29
MgO	7.92	8.36	5.03	2.95	4.22	3.98	6.03
CaO	9.88	8.16	9.72	9.36	10.04	9.88	8.91
Na_2O	3.54	2.36	2.75	4.25	2.24	2.57	3.18
K_2O	2.67	6.18	7.58	2.63	3.57	3.39	1.63
H_2O	2.18	2.60	1.50	2.51	1.74	2.00	1.76
P_2O_5	0.57	0.77	0.50	1.14	0.79	0.84	0.47
(C)*	1.41	1.25	1.30	1.24	1.21	1.20	1.24

*C is a chemical parameter according to BEYERSDORFER

The composition types of basalts are given in Table 1. It appears from the table that the compositions are rather diverse; the "C" value in the table is a chemical composition parameter introduced by BEYERSDORFER [153]. The higher this number the easier will be the rock to melt. In most cases a pure basalt type corresponding to the table is not available, since the rock deposits are mostly intermediaries between the various types. The most readily melting basalt types melt already at 1100 °C, the poorly melting ones above 1200 °C. The strongly basic types have low melting points, the acidic basalt types melt above 1200 °C. Fusibility decreases with increasing silica content. The liquidus temperature of basalts varies between 1150 and 1350 °C. Olivine basalts have high liquidus temperatures due just to their olivine content. The olivine crystals often refuse to melt, but remain in a solid state in the melt. The melting range of the rock between 1350 and 1400 °C ensures a rapid transition into the liquid state and furnishes a homogeneous or almost homogeneous melt.

VOLDÁN's experiments with various basalts have shown that for processing into molten rock basalts containing large quantities – if possible over 60% – of pyroxenes should be used. These ensure abrasion resistance, mechanical strength and chemical resistance. The presence of magnetite and olivine is also important because these will promote crystallization. It is, however, desirable that the quantity of magnetite and olivine shall not exceed 10%, for otherwise the silica content will decrease leading to coarser structures with large crystals and a parallel increase in brittleness.

Large olivine occlusions in basalts are unfavorable, since olivine is a sluggishly melting mineral and under technological conditions coarse, large olivine crystals will refuse to melt to the desirable degree. Rocks with nepheline to plagioclase ratios of 1 : 1 to 1 : 3 and total quantities of approximately 20% are the most favorable. The presence and proportion of nepheline and plagioclase controls the viscosity of the melt and the crystallization process. VOLDÁN assigns the processability of the rock to the composition shown in Table 2.

TABLE 2

Composition of processable
basalts according to VOLDÁN

Oxide	%
SiO_2	43.5—47.0
P_2O_5	0.5— 1.0
Al_2O_3	11.0—13.0
Fe_2O_3	4.0— 7.0
FeO	5.0— 8.0
MnO	0.2— 0.3
TiO_2	2.0— 3.5
CaO	10.0—12.0
MgO	8.0—11.0
Na_2O	2.0— 3.5
K_2O	1.0— 2.0

The petrous products owe their applications primarily to their abrasion resistance. The chemical resistance because of their compositions is not absolutely unobjectionable. In the case of tensile and flexural stresses the application of these materials is not advised because of their rigidity and fluctuating mechanical properties [18, 56, 65, 70].

Deficiencies in strength are due mainly to the already mentioned heterogeneous, coarse crystalline structure. In the course of primary crystallization the structure develops under the influence of a great temperature gradient. In the course of production the surface of the cast products cools rapidly, while inside the body the temperature remains practically the same or may even increase as a result of the liberated crystallization heat. In the case of molten basalt this value is fairly high, about 100 cal per gram. Consequently in the inside of thick products coarse grained, skeleton-like, sometimes idiomorphic crystals may develop. In the direction of the surface this structure changes through dendritic crystals gradually into spherulitic crystal forms. Structural differences due to the thermal gradient, that is the unhomogeneity of the melt, are manifest in the strength of the primarily crystallizing material.

In the case of molten basalts primary crystallization might be suppressed by melting at higher temperatures or by the choice of more acidic compositions, but in this case the method of secondary crystallization cannot be applied because of the deformation of the material. This points to the importance of the nucleating agents used in the production of vitroceramics.

(B) Conditions for the formation of vitroceramics

Certain silicate chemical problems whose study is necessary for developmental work, that is for the determination of optimum preparation conditions, can be determined from the above outlined process of thermal treatment of vitroceramics.

It is a known fact that the crystallization of glass begins mainly at the interface as a result of heterogeneous nucleus formation. Without nucleating agent this will cause crystallization to proceed from the surface towards the inside of the glass body.

Nuclei are easily formed on the surface of the glass sample and when the nucleus has reached the size of a crystal, crystals begin to grow on the surface. The crystalline surface layer tends to shrink, but shrinkage is hindered by the internal, still amorphous, mass. If the crystallization of such a glass proceeds, the surface will be crimped once crystallization begins to progress in the internal layers. This phenomenon is caused by the contraction accompanying crystallization. If crystallization starts from the surface, in the inside of glass rods longitudinal internal voids may be formed and the rod will become a tube.

The formation of vitroceramics and the nature of their structure depend on several physical-chemical parameters. Since TAMMANN the crystallization of liquids is characterized by the crystallization ability which is the sum of two factors [76–78]:

(a) crystallization rate and
(b) crystallization tendency.

The correlation between these two factors is shown in Fig. 6. Crystallization rate is of a definitely kinetical nature, while crystallization tendency is characterized by the distribution, that is, frequency of crystalline nuclei in space. Crystallization ability is further characterized by the extent of the crystallization area, that is, by the difference between the temperatures of the lower and upper limits of crystallization. In accordance with the kinetical situation the crystallization process is treated by starting from the liquid state in the direction of decreasing temperature. Thus the crystallization area begins at the liquidus temperature (T_L) and ends at temperature T_A that is where crystallization rate (v) and crystallization tendency (N) approach the zero value. Both crystallization rate and crystallization tendency are known to have maxima between T_L and T_A at the temperatures $T_{v\,max}$ and $T_{N\,max}$ respectively. In

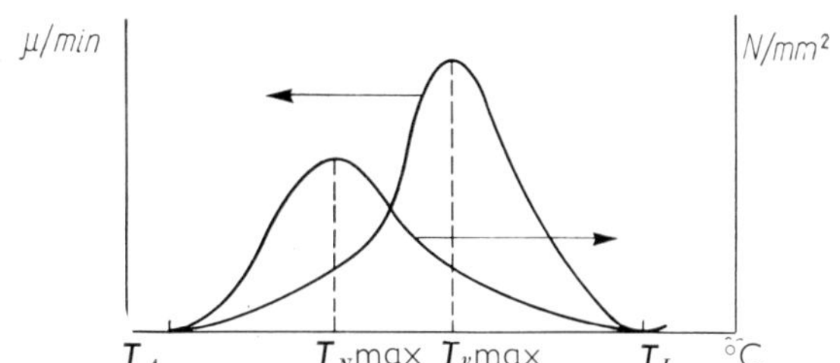

FIG. 6. Changes in the crystallization ability of glass according to TAMMANN

the case of silicate melts $T_{v\,max} > T_{N\,max}$. Crystallization rate maximum is at a higher temperature than the maximum of the number of crystallization nuclei. The difference between the temperatures pertaining to the two maxima is also characteristic of the crystallization process. This difference has a particular importance from the aspect of the preparation of vitroceramics, since this determines the kinetical nature of secondary crystallization leading to a very finely dispersed crystal structure.

Crystallization tendency plays the following role: the condition for the formation of vitroceramics includes the melts first passing the crystallization temperature zone without primary crystallization in the amorphous phase. The composition range which form the aspect of the production of vitroceramics is of any interest may be limited on the basis of the crystallization tendency of the silicate melt. If the crystallization temperature zone is approached in a secondary manner from below, then the crystallization process will be characterized by a high number of crystallization nuclei and a low crystallization rate as shown in Fig. 6 [79–81].

Crystallization tendency limits the useful compositions. Special stable glasses are unsuited in the same manner as those with too high crystallization tendencies for vitroceramic base glasses. In the first case crystallization time will be too long, in the second case secondary crystallization will not be established because of the commencement of primary crystallization.

In the production of vitroceramics the conditions of the formation of the vitreous state enter the considerations with almost the same weight as in the course of glass production. Hence it is necessary for the nucleation activation energy to exceed the level otherwise required for the initiation of the primary crystallization process and for attaining a low liquidus temperature.

Shaping operations are similar partly to those of glass production and partly to iron and steel casting. The amorphous phase and the existence of an undercooled liquid are here too first of all the functions of the cooling rate. Thus cooling rate plays an important role in the development of the composition of the vitroceramic melt, and this rate is determined by the well-known conditions of shaping methods.

Cooling rate is an important kinetical factor in glass formation with which we have to reckon here too and together with the glass composition the parameters of the kinetical processes, such as heat capacity and heat transfer conditions, must also be given. This is of particular importance in the case of new glass compositions.

The importance of the liquidus temperature is the same for the production of vitroceramics as of glass. The thermodynamical content of the liquidus temperature is in both cases decisive from the aspect of shaping: above this temperature no nucleation occurs. The average kinetical energy of the molecules in the silicate liquid is according to the statistical distribution of ARRHENIUS proportional to the value of

$$\mathrm{Exp} - \frac{E}{RT}.$$

The rate of the kinetical process of crystallization is:

$$V = Z \, \mathrm{Exp} - \frac{E}{RT}$$

that is a function of the absolute temperature and of the energy threshold E.

Chapter IV

The governed crystallization
of glasses

According to the modern theories on vitreous structures an aggregation
process will begin during the cooling of a liquid. This aggregation is caused
by the same forces which under normal conditions lead to the formation
of crystals of regular structure. Recent theories suppose the formation
of an intermediary structure between the fully ordered and fully irregular
states during the cooling process (VOGEL, PREBUS and MICHENER, SELIUB-
SKII, OBERLIES, PORAI-KOSHITS, HOFFMANN and STATTON [82–92]). The
appearance of occasional more ordered structural elements may be considered the preliminaries of crystallization. There is a fairly great difference
between the energy contents of truly amorphous and "pre-ordered" parts
of the solidifying structure. This state carries within it the possibility of
a separation of the parts with different energy contents by well-defined
interfaces. This difference is, however, yet remote from indicating a disrupted
amorphous structure of the glass, but is very close to the occurrence of
a well-defined microphase separation.

(A) Homogeneous and heterogeneous nucleus formation

The definition of glass includes a lack of crystallization when the material
is cooled from the liquid state under definite kinetical conditions. The possibility of crystallization arises however as soon as the kinetical conditions
of glass formation are not fulfilled. The value of the kinetical condition,
which can be defined in the $(T_L - T_A)$ interval by the cooling rate, is
determined by the energy of nucleus formation and the product

$$\eta(T_L - \Delta T)$$

in the given interval. The higher the energy of nucleus formation the more
favorable will be the conditions for glass formation and the vitreous state
will be the more stable.

Since the valency bonds of the ions forming the surface structural elements are balanced only in the direction of the glass inside, the energy of
nucleus formation will be reduced by the surface free energy so that nucleus
formation, that is subsequent crystallization, will begin on the surface.
On the surface the presence of foreign substances may interfere, so that
this phenomenon may also be considered a type of heterogeneous nucleus
formation.

Nucleus formation is considered homogeneous if the associated surface
formation is brought about without a foreign phase, that is without inter-

face. This means that an equal possibility exists in every point of the melt for homogeneous nucleus formation. According to VOLMER [110] there is an exponential relationship between homogeneous nucleus formation (J) and the quotient of the activation energy of nucleus formation $(-F^*)$ and the absolute temperature:

$$J = A \, \mathrm{Exp} - \frac{F^*}{KT}$$

where K is the BOLTZMANN constant $(R/N_A = R/N^*)$ and A is a constant.

According to TURNBULL the following relationship is valid for heterogeneous nucleus formation:

$$I = A' \, \mathrm{Exp} - \frac{F^* f\theta}{KT}$$

where I is the rate of nucleus formation on a catalyst surface. F^* is the maximum activation energy required for nucleus formation on a catalyst surface, $f\theta$ is the energy decrease on the surface of a liquid wetting the catalyst surface and θ is the contact angle of the liquid on the catalyst surface.

The effect of the diffusion rate on crystallization in the course of nucleus formation is expressed by the following formula:

$$I = A' \, \mathrm{Exp} - \frac{F^* f\theta + q}{KT}$$

where q is the diffusion term.

The activation energy of heterogeneous nucleus formation is considerably lower than of homogeneous nucleus formation which is manifest e.g. in the beginning of crystallization on the surface of a glass containing a highly dispersed foreign phase (for instance metal) at temperatures 50–100 °C lower than on the surface of such glasses of otherwise identical composition, but without nucleating agent. Crystallization in the first case will begin on the surface of the foreign phase. The nucleating agent may separate from the liquid by homogeneous catalysis provided primary micro-regions due to phase separation have been formed. VOGEL [94, 103–106] and STOOKEY [95] consider this state a very favourable pre-crystallization state. VOGEL and GERTH [96] claim that the separation phenomenon may be promoted by modifying the following factors:

1. the relative concentrations of the glass components
2. heat treatment
3. reduction of the field intensity of the cations of the glass
4. application of small quantities of additives.

There are a number of cases when in the interest of producing vitroceramics the activation energy of phase separation should not be decreased but increased in order to be able to cool a given composition in its amorphous state. In certain cases to ensure strictly defined physical chemical properties a composition has to be chosen which cannot be cooled without preliminary crystallization [97–102]. LŐCSEI [108, 109] has proved for a number of cases the reduction of primary crystallization tendency, that is the increase of the activation energy required for phase separation, by a method which he calls the "micro-eutectic principle".

The stability of the vitreous phase as manifest in the reduced separation tendency of the submicro-regions of phase separation can be best described by the ratio between the network forming and modifying ions, that is with the quantity of oxygen ions in bridge position. It is further known that partly polarization due to the nature of the modifying ions and partly their quantity may influence the stability of the vitreous state which is essentially in agreement with the findings of VOGEL and GERTH [93] applying these in a negative sense to pars 1, 2 and 3.

The effect which may be achieved by raising the number of components may be interpreted firstly by a qualitative and only secondly by a quantitative change. LŐCSEI [106, 108] has shown in his experiments that the stability of the vitreous state may be increased by the incorporation of relatively low quantities of new components provided there are many components. He confirmed the validity of this statement by the investigation of a model glass of the following composition:

$$SiO_2 \qquad 67\ \text{mole}\%$$
$$Na_2O \qquad 18\ \text{mole}\%$$
$$CaO \qquad 15\ \text{mole}\%$$

This is a ternary glass composition which in the sense of the WARREN–ZACHARIASEN theory represents a border case [111–114] of the criterion of stable vitreous state formation on the basis of the quantity of modifying ions. LŐCSEI introduced into this composition new components in 1 mole% quantities and raised the number of components (N) by 1, 2, 4, 8, 12 and 16, respectively. Substitution was at the expense of silica. In spite of the decrease in the ratio of network forming and modifying ions in the glass that is, in the quantity of oxygen ions in bridge position, the stability of the vitreous state increased – with the exception of the first two substitution methods – with the number of components.

In substitution components 1 and 4 were chosen so that the new components should if possible not play the role of network forming or intermediary ions in the glass structure and their ionic radius shall approximate that of Na^+ and Ca^{2+} respectively. If an odd number of new components is introduced then the ion with the smaller radius shall predominate, while in the case of an even number of substituting ions their proportion shall be about equal. When the number of substituting components is high, it may be necessary to apply network forming and modifying ions too.

This principle may be applied to raise the number of possible compositions for vitroceramics and to improve the possibility of the formation of a favorable microstructure.

Lőcsei also investigated a base glass of the following composition:

SiO_2	69 mole%
BaO	8 mole%
Li_2O	23 mole%

The substituting oxides were in order:

   I. CaO, PbO,
  II. CaO, PbO, SrO, CdO,
 III. CaO, PbO, SrO, CdO, B_2O_3, ZrO_2, ZnO, MgO,
 IV. CaO, PbO, SrO, CdO, B_2O_3, ZrO_2, ZnO, MgO, Al_2O_3, Sb_2O_3, Bi_2O_3, As_2O_3.

The quantity of the substituting oxides was always 1 mole%. Substitution was performed by one of two ways: either at the expense of the silica or of the barium oxide content, that is, the quantity of one of these constituents was reduced by 1 mole%.

Substitution resulted in the following changes in the heat expansion coefficients of glasses: heat expansion coefficient rose from 61.4×10^{-7} mm/mm °C to 67.2×10^{-7} mm/mm °C when substitution was at the expense of the silica content, and to 62.5×10^{-7} mm/mm °C when substitution involved a reduction of the barium oxide content. The densities of the glasses underwent similar changes: in the first case there was an increase by four units in the third decimal of the density, while the second method of substitution had practically no effect on the density. Chemical resistance improved considerably in both cases.

Hydrolysis resistance tests showed a decrease of the dissolved lithium oxide per 1 g of glass from 5.4×10^{-4} mg of Li_2O/g in the first case to 2.3×10^{-4} and to 1.7×10^{-4} mg of Li_2O/g in the second case. Table 3 shows the changes in the stability of the vitreous state and in the extent of the crystalline region. Both types of substitution resulted in a decrease of the crystallization interval. This decrease is about 50% in the first and nearly 90% in the second case.

TABLE 3

Changes in the liquidus temperature and crystallization area
of the lithium barium silicate glass as an effect
of increased composition variability

<table>
<thead>
<tr><th rowspan="2">Number
of components
in the additive</th><th colspan="2">Lower limit
of crystallization
(°C)</th><th colspan="2">Liquidus temperature
(°C)</th><th colspan="2">Crystallization
temperature interval</th></tr>
<tr><th>I</th><th>II</th><th>I</th><th>II</th><th>I</th><th>II</th></tr>
<tr><th>n</th><th colspan="6"></th></tr>
</thead>
<tbody>
<tr><td>0</td><td>—</td><td colspan="... ">930</td><td>—</td><td colspan="...">590</td><td>—</td><td colspan="...">340</td></tr>
</tbody>
</table>

Number of components in the additive (n)	Lower limit of crystallization (°C) — I	II	Liquidus temperature (°C) — I	II	Crystallization temperature interval — I	II
0	—	930	—	590	—	340
2	920	910	600	610	320	300
4	900	870	620	650	280	220
8	860	830	640	700	220	130
12	840	810	660	770	160	40

Figure 7 shows the crystallization pattern of the base glass after 3-minute heat treatment at 840 °C, Fig. 8 is the pattern of the glass with twelve additive components after 30-minute heat treatment at 840 °C. Comparison of the two patterns shows a decrease of the crystallization rate to about one-tenth of the rate in the base glass [109].

In another series of experiments Lőcsei intended to prove that the greater variability of the glass composition results partly in a significant reduction of the crystallization ability and partly in a considerable increase in the number of crystallization nuclei. In this series of experiments the base glass belonged into the SiO_2–Al_2O_3–CaO–MgO–Na_2O system which is one of the vitroceramic base glasses worked out by Lőcsei [22, 23] by the substitution of 3–4% of Na_2O by 3–4% manganese and iron sulphides.

SiO_2	60%
Al_2O_3	10%
CaO	15%
MgO	7%
Na_2O	8%

The total quantity of the additives was composed of BaO, SrO, ZnO, CdO, PbO, TiO_2, ZrO_2, Sb_2O_3, B_2O_3, SnO_2; substitution was at the expense of the CaO and MgO contents. Crystallization rate and the number of nuclei were determined on cooled samples of the glasses. The crystallization temperature interval was approached from the direction of lower temperatures.

Figure 9 shows the crystallization of the base glass after 4-hour crystallization at 800 °C ($100\times$ magnification).

The crystallization state of the glass with additives is shown in Fig. 10 after 1-hour heat treatment at 800 °C ($500\times$ magnification). Crystallization of the glasses both with and without additives begins at the surface. Comparison of Figs 9 and 10 shows clearly (especially when we account for the difference in magnifications) that the crystallization tendency of the glass without additive is approximately 25 times higher than that of the glass with additive. (One division in Fig. 9 corresponds to 10 μ, in Fig. 10 to 2 μ.)

Figure 11 is the crystallization pattern of the glass with additive after 1-hour crystallization at 850 °C ($100\times$ magnification), while Fig. 12 represents the crystallization of the glass without additives under identical conditions. (In both Figs 11 and 12 one division corresponds to 10 μ.) It appears from the comparison of the figures that at the given temperature the crystallization rate of the glass with additives is much lower than of the glass without additive, while the number of crystals in the glass with additive is several times higher than in the base glass.

The effect of additives after heat treatment at 900 °C for 1 hour is presented by Fig. 13 ($100\times$ magnification); Fig. 14 shows the glass without additive after crystallization at 900 °C for 30 minutes ($100\times$ magnification).

Figure 15 is the picture of the crystallization conditions of the glass with additive after heat treatment at 950 °C for 1 hour ($100\times$ magnification),

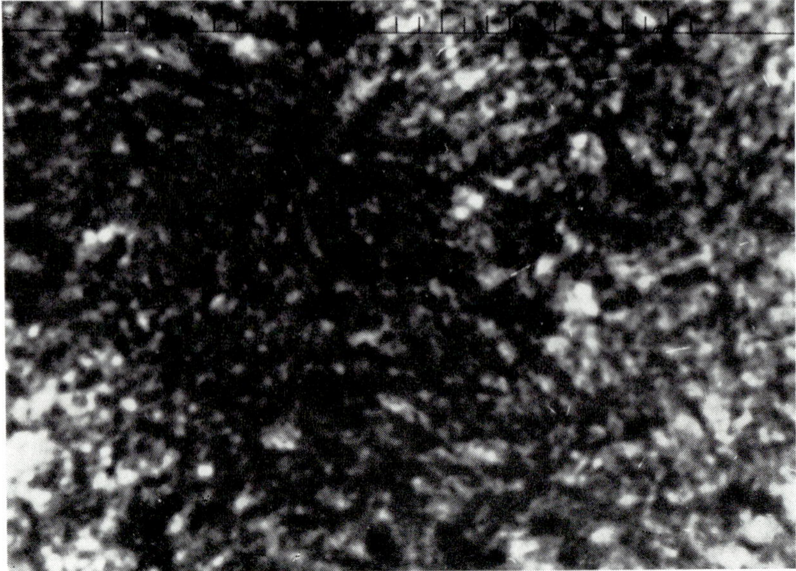

FIG. 7. Crystallization of lithium barium silicate glass at 840 °C under the effect of 3-minute heat treatment

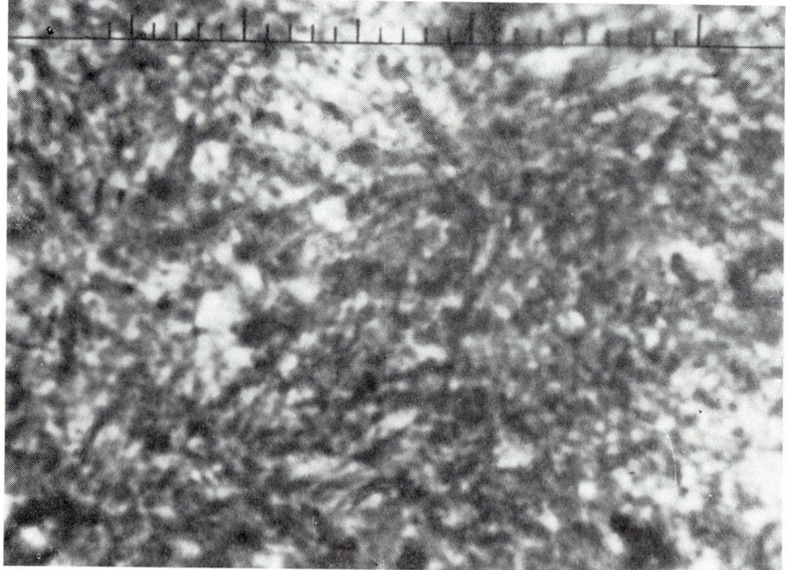

FIG. 8. Crystallization of a lithium barium silicate glass containing 12 additive components under the effect of 3-minute heat treatment at 840 °C

FIG. 9. Surface crystallization of a base glass in the system $SiO_2–Al_2O_3–CaO–MgO–Na_2O$ at 800 °C in 10 hours. Magnification: $100\times$

FIG. 10. Surface crystallization of a base glass in the system $SiO_2–Al_2O_3–CaO–MgO–Na_2O$ containing 1% of 10 different additives at 800 °C in 10 hours. Magnification: $500\times$ (5 times that of the glass without additive)

Fig. 16 that of the glass without additives after heat treatment at the same temperature for 30 minutes (100× magnification). Because of the great number of nuclei and the low crystallization rate of glasses with additives no large crystals will appear, while in glasses without additives because of the smaller number of nuclei large acicular crystals will develop within the same or shorter period of time. In Figs 13–16 one division corresponds to 10 μ.

Investigation of the crystallization process points unequivocally to an increase in the number of nuclei and a significant decrease in crystallization rate as a result of the presence of additives. Identical or shorter heat treatment periods bring about in glasses without additives crystals several times larger than those observed in the glass with additives which has been subjected to the same heat treatment [106, 109].

Figures 17 and 18 illustrate the phase separation phenomenon which occurs as a result of heat treating a glass in the SiO_2–Al_2O_3–CaO–MgO–Na_2O system and prepared by means of sulphide nucleation [1, 20, 79, 81, 100].

At 780 °C MnS and FeS separate in the glass belonging to the five-component system. Its state prior to heat treatment is shown in Fig. 17 (about 54,000× magnification). The microscopic picture (Fig. 18) taken after heat treatment at 780 °C for 15 minutes proves quite clearly the separation of droplets [106]. As demonstrated by Lőcsei [22, 100] for sulphide nucleation, the occurrence or absence of the separation phenomenon, that is, its primary or secondary appearance, depends on the temperature, heat capacity, heat transfer and mass of the glass. These factors will determine – beside the shape of the glass – the rate of cooling. Figure 19 shows the separation of phases in a larger body. There will be no primary phase separation when during the cooling of an approximately spherical glass body of 20 mm diameter the temperature difference (ΔT) is 1200 °C, but when ΔT is 700 °C, that is the glass sphere cools in a space of 500 °C temperature, primary phase separation phenomenon, such as illustrated in Fig. 19, will occur [171].

From the aspect of crystallization mechanism the number of separated heterogeneous drop-shaped nuclei is an important factor, since this determines the formation of the vitroceramic material structure. Frenkel's thermodynamical deduction [107] is applicable also to the processes of nucleus formation and phase separation. Phase transition and the formation of a surface separating the new and the old, that is the crystalline and the liquid phase, are required for the occurrence of the separation phenomena illustrated in Figs 18 and 19. The local increase of free energy enabling the occurrence of this phenomenon is furnished by the vibration of the liquid system.

The kinetics of homogeneous nucleus formation is characterized partly by the increase of the free energy and the velocity of atom transport necessary for its development. The free energy increase includes the energy needed for the formation of the new surface. Below the liquidus temperature from the unstable liquid with higher internal energy content thermodynamically stable, submicroscopic crystal nuclei are formed.

Fig. 11. Surface crystallization of a base glass in the system SiO_2–Al_2O_3–CaO–MgO–Na_2O containing 1% of 10 different additives at 850 °C in 10 hours. Magnification: 100×

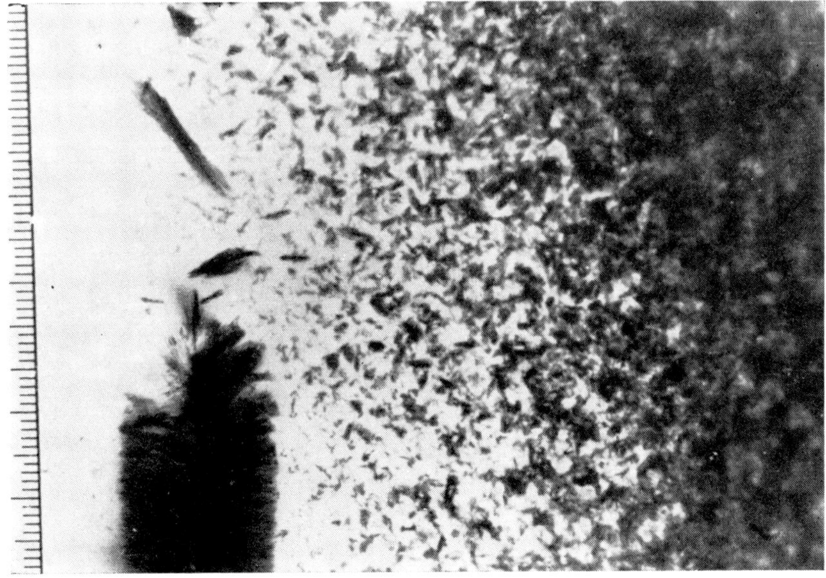

Fig. 12. Surface crystallization of a glass in the system SiO_2–Al_2O_3–CaO–MgO–Na_2O without additive at 850 °C in 1 hour. Magnification: 100×

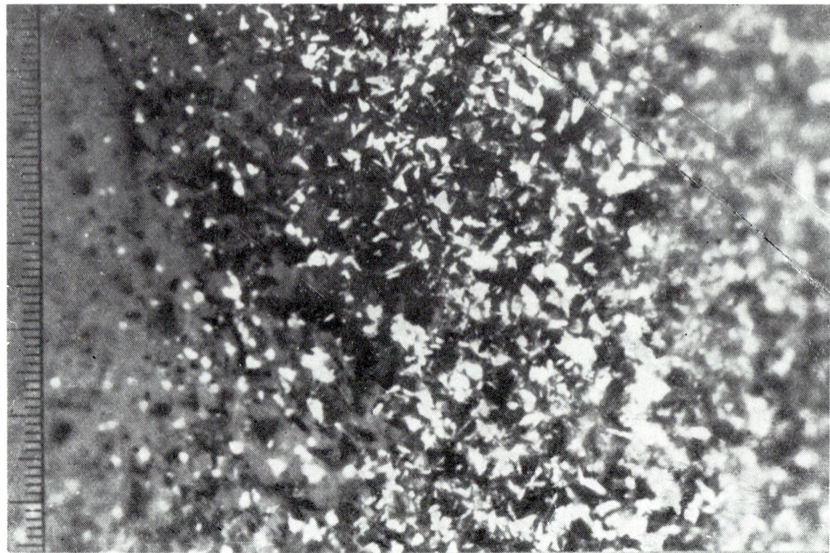

FIG. 13. Crystallization of a glass in the system $SiO_2–Al_2O_3–CaO–MgO–Na_2O$ containing 1% of 10 different additives at 900 °C in 1 hour. Magnification: 100×

FIG. 14. Surface crystallization of a glass in the system $SiO_2–Al_2O_3–CaO–Mg–ONa_2O$ without additives at 900 °C in 30 minutes. Magnification: 100×. Crystallization time was only half of that applied to the glass with additives and the crystals are nevertheless 2–3 times larger

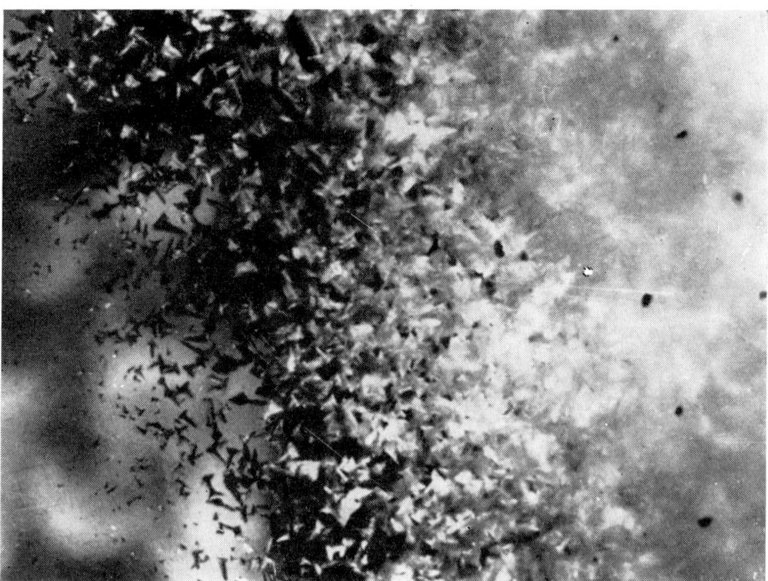

FIG. 15. Crystallization of the glass in the system SiO_2–Al_2O_3–CaO–MgO–Na_2O containing 1% of 10 different additives at 950 °C in 1 hour. Magnification: 100×

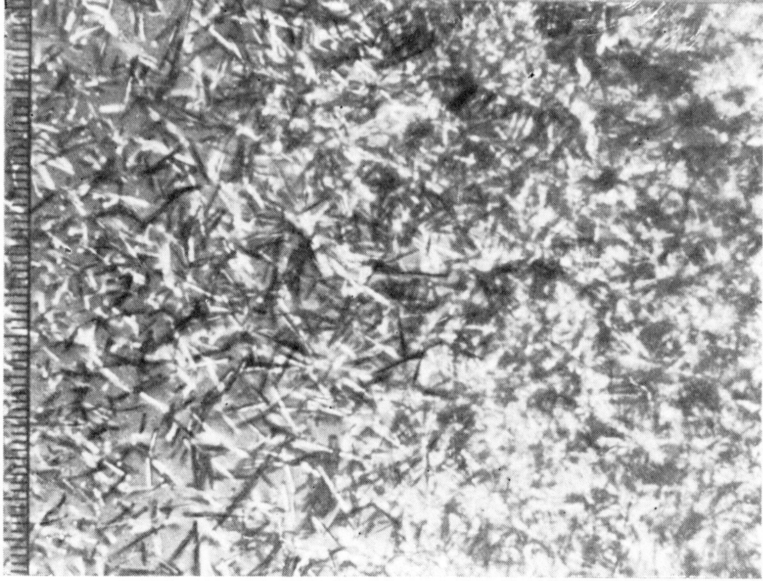

FIG. 16. Surface crystallization of a glass in the system SiO_2–Al_2O_3–CaO–MgO–Na_2O without additive at 950 °C in 30 minutes. Crystallization rate is 6 times higher than in the glass with additive

FIG. 17. A sulphide containing vitroceramic base glass cooled in the X-ray amorphous state. Magnification: 54,000×

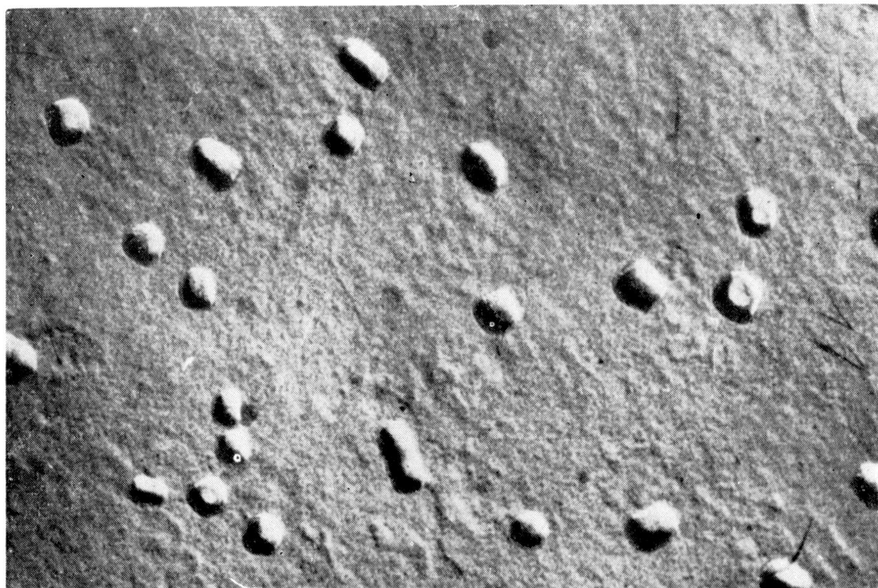

FIG. 18. Droplet-like separation phenomenon in the X-ray amorphous glass after heat treatment at 700 °C for 15 minutes. Magnification: 54,000×. (The electron microscopic pictures were taken in the Electron Microscopic Laboratory of the Hungarian Academy of Sciences)

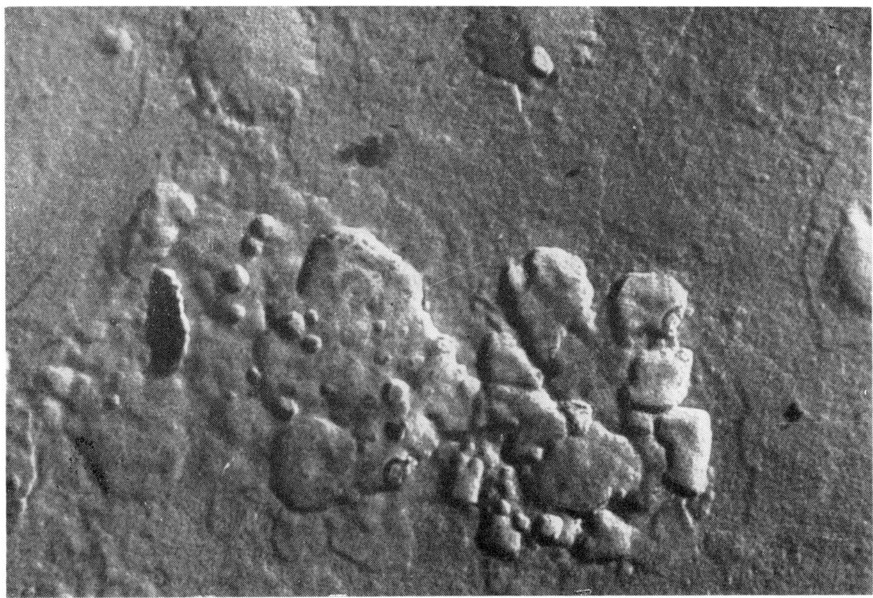

FIG. 19. Phase separation phenomenon in a slowly cooling vitroceramic material with sulphide content. Magnification: 54,000×

If the radius of the formed crystal nucleus is r and the surface tension of the melt σ, then the thermodynamical potential change of the system for the crystallization process may be written as follows:

$$\Delta\Phi = \Phi - \Phi_0 = \Phi_A N_A + \Phi_B N_B + 4\pi r^2 \sigma - \Phi_A(N_A + N_B) \tag{1}$$

where Φ is the thermodynamical potential of the system which contains N_A particles in the liquid and N_B particles in the crystalline state; Φ_0 is the thermodynamical potential of the liquid system which contains $N_A + N_B$ elementary particles; Φ_A is the chemical potential of the system under investigation in the liquid state and Φ_B its chemical potential in the solid state.

If the volume of the approximately spherical elementary particles is taken as

$$\frac{4r^3\pi}{3N_B} = V_B$$

then Eq. (1) may be written also in the form:

$$\Delta\Phi = \frac{\Phi_A - \Phi_B}{N_B}\frac{4\pi r^3}{3} + 4\pi r^2 \sigma. \tag{2}$$

The condition of thermodynamical equilibrium is

$$\delta(\Delta\Phi) = 0.$$

3*

Thus we may write that

$$\frac{\Phi_A - \Phi_B}{V_B} = \frac{2\sigma}{r'} \tag{3}$$

where in the case of thermodynamical equilibrium $r = r'$. Hence

$$\Delta\Phi = 4\pi\sigma\left(-\frac{2r^3}{3r'} - r^2\right). \tag{4}$$

The thermodynamical potential ($\Delta\Phi$) increases when $r \rightarrow r'$, once it reaches the value pertaining to r' it will decrease according to Fig. 20.

The maximum thermodynamical potential is:

$$\Delta\Phi_{max} = \frac{4}{3}\pi r'^2 \sigma.$$

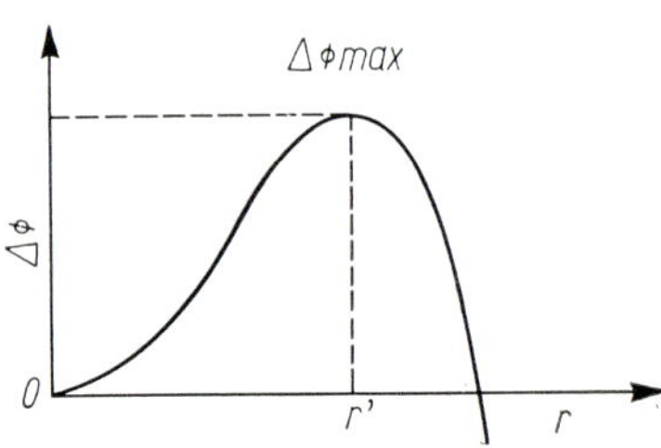

FIG. 20. The thermodynamical potential ($\Delta\Phi$) vs. r. The thermodynamical potential has a maximum at $r = r'$

It follows from the statistical probability of the melt structure and the value of the thermodynamic potential that in a cooling liquid the probability of nucleus formation will be the same at any point of the liquid. $\Delta\Phi_{max}$ is the activation energy of crystal formation. Nuclei which have reached the critical dimension r' will grow at a rate depending on the crystallization rate and the process is controlled by the viscosity of the system and the diffusion rate [123–127].

Processes taking place in the formation of the following glasses may be considered the results of homogeneous nucleation:

1. opal glasses (fluoride, phosphate)
2. ruby glasses (Au, CdSSe, Cu)
3. aventurine glasses (chromium oxide)
4. enamels (TiO_2, ZrO_2).

Figure 21 shows the thin section of a fluoride opal glass in transmitted light (about $1800\times$ magnification). The separated phase was roentgenographically identified as consisting of CaF_2 and NaF. Figure 22 is a microscopic picture of a "haematinone" glass section prepared by VERESS [122] in transmitted light. The spots are separated metallic copper, in the middle a larger sphere is visible with some tetrahedral crystals here and there, indicating that copper oxide was reduced around its melting point, or at a temperature not much exceeding its melting point to metallic copper. Copper was separated in a highly viscous state of the melt, otherwise large spherical agglomerations (such as appears in the middle of the picture, or even larger) would have been formed exceeding the dimension at which a suspension can occur. In this case the separated crystals sink in the glass.

Figure 23 shows a thin section of a selenium ruby glass. The glass is transparent and red. The dark spots in the figure appear orange colored

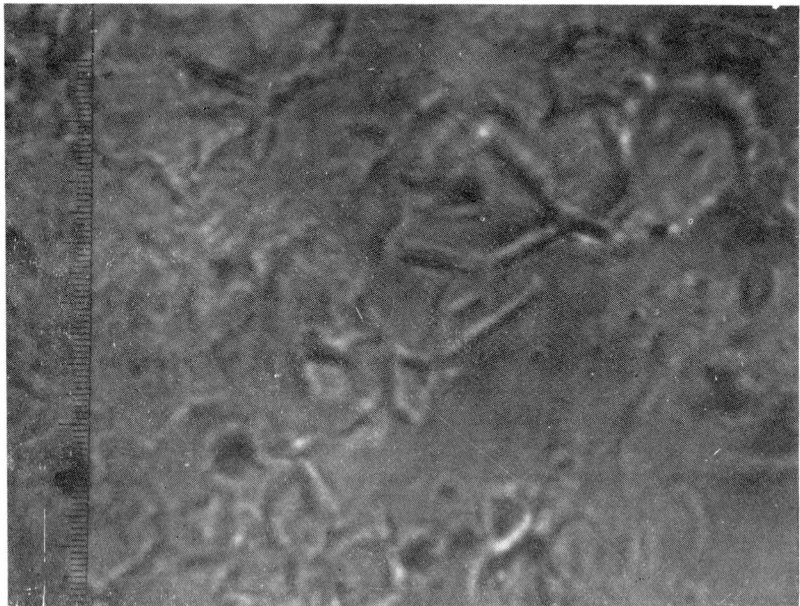

FIG. 21. Thin section of an opal glass. Magnification: 1800 ×

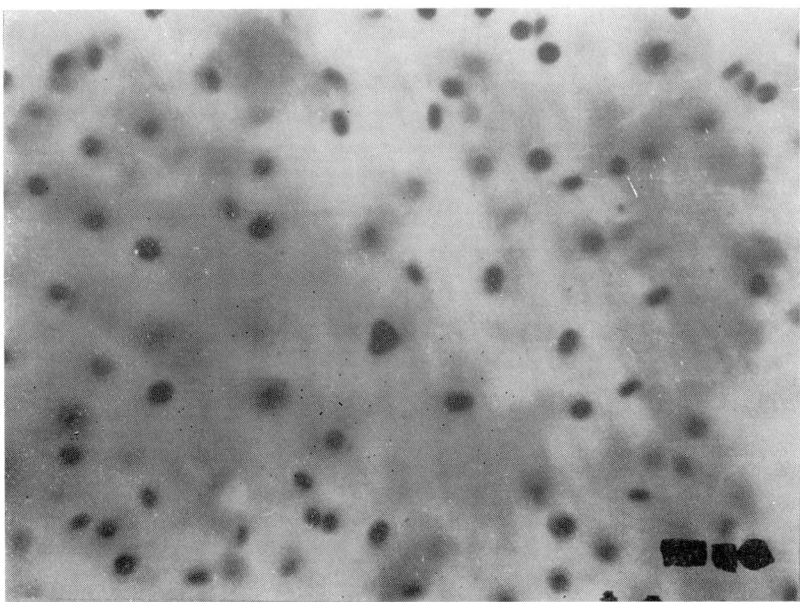

FIG. 22. Thin section of a haematinone glass in transmitted light showing the separation of metallic copper

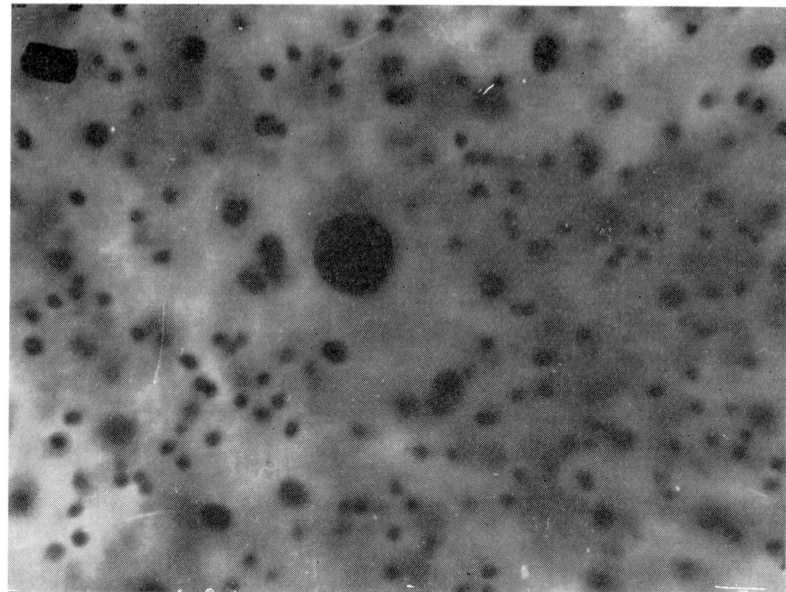

Fig. 23. Slowly cooled selenium ruby glass section in transmitted light showing the separation of 0.2–1.0 μm cadmium selenide crystals

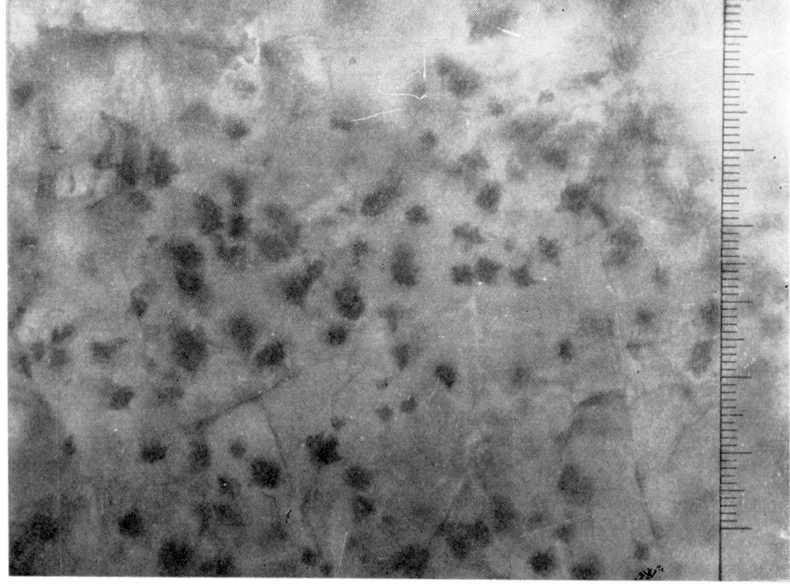

Fig. 24. Thin section of titanium enamel. Magnification: 400×

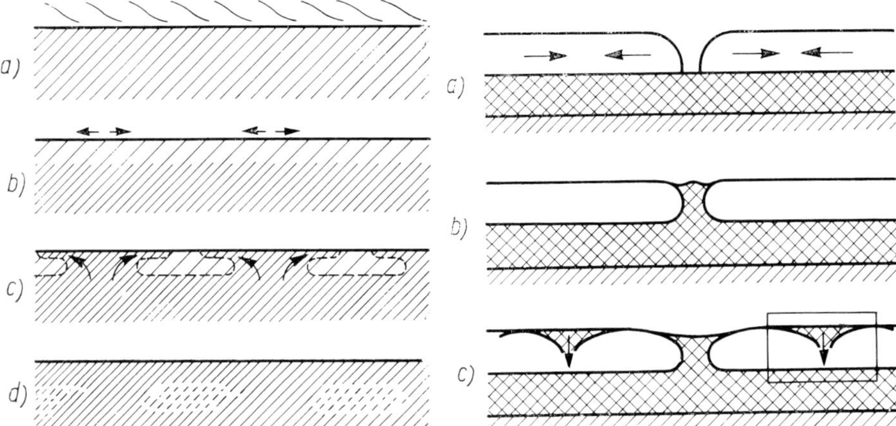

Fig. 25. The Jebsen–Marwedel diagram of separation in glass under the action of dynactivity. The effect of reducing furnace atmosphere on the liquid state: a) the effect of reducing atmosphere on SO_3 containing glass; b) separation of the surface layer with reduced SO_3 content; c) next step in the separation process; d) sinking of the glass with reduced SO_3 content to the lower layers

Fig. 26. Behavior of two enamel types of different surface tensions. Penetration of the primer enamel into the covering coating: a) due to its high surface tension the covering enamel breaks; b) the primer enamel penetrates the crack in the coating; c) the primer enamel now on the surface penetrates again from above towards the first coating

under the microscope. They are ditrigonal crystals with the interesting feature of being longer in the direction of their side-axis (0.3–1.0 μm).

Figure 24 is the thin section of a titanium enamel. These phenomena based on phase separation are applied in glass and enamel industrial practice and are the results of the growth of crystal nuclei formed by homogeneous nucleation. This phase separation phenomenon occurs only under special conditions and is the consequence partly of the glass composition and partly of the fluoride, chromium oxide, copper oxide and titanium oxide content. In this case only the so-called additives have yet crystallized, not so the vitreous silicates. If the glass composition is appropriate after crystal, that is metal separation crystallization can be propagated by heat treatment to reach the vitreous material. The new phase formed by phase separation due to homogeneous nucleation will now act as a nucleating agent. Vogel [114] supposes from the experiments of Jebsen–Marwedel [115–121] a correlation between the "dynactivity" of silicate melts and the phase separation phenomenon of glasses. The mechanism of macroscopic separation in silicate melts as shown diagrammatically in Figs 25 and 26 may be traced back to chemical and physical differences.

(B) Nucleation and nucleating agents

St. Claire Devill used the term "mineralizer" which was then applied by Fouqué and Michel-Lévy generally to materials accelerating geochemical processes [128, 129]. Though the term has been adopted later by physical chemistry, in the Anglo-Saxon literature instead of mineralization, nucleation is the accepted name. The formulation of generally valid rules began from the evaluation of experimental results mainly of technological nature. The importance of nucleation increased with the development of solid phase reaction kinetics. The reaction rate of cement hardening is accelerated by the fluorides of alkaline earth metals. Mullite formation is very favorably influenced by decreasing cation and halogen ion radii [130–139]. The rate of polymorphous transformation, especially in silica and alumina systems, may be beneficially influenced by certain nucleating agents (mineralizers) [140]. The negative, passivating effect of certain mineralizers has also been demonstrated [141]. The formation of solid solutions in the course of several solid phase reactions may have a stabilizing effect, as for instance in the case of ZrO_2 or dicalcium silicate.

In the acceleration of sintering processes the nucleating agent displays its action by promoting the formation of a melt phase or recrystallization. In solid phase reactions the formation of intermediary phases are of special importance, as e.g. the amorphous and crystalline aluminum fluosilicate in the kaolinite–AlF_3 system [142]. One of the important features of nucleation has been shown by the experiments to consist in the reduction of the surface tension of the melt phase. This type of effect was found in mullite synthesis, or during the sintering of enstatite, dolomite, etc. [143–147].

Avgustinik ascribed the effect of nucleating agents in general to their ability to reduce the energy necessary for the formation of new phases.

The importance of nucleating agents is perhaps the greatest in three fields:

1. sintering
2. formation of the liquid phase
3. formation of the crystalline phase from the melt.

The first effect, namely sintering may be described by three processes:

(a) the diffusion of the ions on the surface;
(b) mutual diffusion in space of the ions of the base phase and of the nucleating agent according to the kinetics of Frenkel's void mechanism;
(c) displacement of ions or ion groups without activation energy in the direction of the Buerger vector; perpendicularly to this direction the activation energy has a maximum.

The second effect, that is acceleration of the formation of the liquid phase, is essentially manifest in the nucleating agent promoting the displacement of the ions of which the crystal is built from their strictly defined steric position, thus the increase of the energy content of the system.

The third effect is manifest in an enhanced crystallization of the liquid, and the nucleation effect appears in a reduced activation enthalpy requirement for the formation of the new phase [129].

The first step in the understanding of the kinetical processes is the clarification of the qualitative and quantitative role of the nucleating agents. This is of particular importance from the aspect of both the theory and practice of vitroceramics.

In the production of vitroceramics the main effect of nucleating agent addition is firstly that these substances crystallize primarily from the amorphous mass during heat treatment and fulfil the role of crystallization nuclei. From the aspect of the process it is highly characteristic and important that the nucleating agent additive separates at a temperature which is lower than the crystallization limit of the base glass. The new phase, especially if crystalline, raises the viscosity of the heterogeneous system and reduces the deformation danger due to crystallization. Because of the high specific surface of the separated nucleator the newly formed interface is very great. Due to the high surface energy of the glass in contact with the nucleating agent crystallization will begin under the influence of heat treatment earlier in the inside of the mass, at the border of the new phase formed by the nucleating agent, than at the glass surface. Thus in fact crystallization will proceed from the inside towards the outside and consequently the phenomena which occur in the case of glasses crystallizing without nucleation will be absent. The surface will not crease, internal voids and cracks will not be formed, there will be no deformation.

Crystallization of the so-called base glass remaining after the separation of the nucleating agent will be determined in the following by the crystallization ability of the system, by the number of crystal nuclei, by crystallization rate and liquidus temperature and by the crystallization area which can be characterized by the lower temperature limit. The nucleating agent which is present in a finely dispersed state in the glass body will reduce the individual crystal sizes by forming crystallization centres. The structure will be similarly influenced by the number of crystallization nuclei and crystallization rate of the base glass. The lower the crystallization rate and the higher the crystallization tendency of the base glass, that is the greater the number of the formed nuclei, the smaller will be the individual crystals of which the structure is made up which offers first of all a possibility of evolving highly favorable mechanical properties. No crystallization must occur in the cooling stage of the shaping of vitroceramics; the material must be cooled in the amorphous state below the lower limit of the crystallization area.

There are three ways for further development:

1. Full investigation of the quality and intensity of the nucleation effect. Certain conditions affecting the functioning of ions of nucleating action, e.g. inhibitor effects have to be clarified, since certain substances with nucleating action will display their effect only depending on the composition of the base glass.

The specific action of the nucleators is manifest in the same nucleator displaying a favorable action on certain composition types, while being completely ineffective on others. This phenomenon may be eliminated by nucleator combinations, that is the combined application of several nucleating agents.

2. Investigation of the crystallization ability of the base glass as a structure modifying factor. These investigations may furnish valuable data not only from the aspect of influencing the structure of vitroceramics, but the same data may be utilized to reduce the crystallization tendency of industrial glasses.

3. Control of the liquidus temperature of vitroceramics has a great theoretical and practical importance. Increase of the liquidus temperature was obviously solved by raising the total SiO_2 and Al_2O_3 content, while for its reduction the separate or combined application of B_2O_3, V_2O_5, Sb_2O_3, PbO and P_2O_5 is suggested. Combined with the control of the heat expansion coefficient important results may be achieved with respect to soldering to metals or the application of vitroceramics to metals [148].

In his study of the nucleation process and the formation of the vitreous state WEYL found that the following factors will determine the kinetical situation [149]:

1. cooling rate
2. the liquidus temperature (T_L) and the lower limit of the crystallization area (T_A)
3. parameters determining the correlation between viscous flow and the activation energy of nucleation
4. parameters determining the correlation between viscous flow and the activation energy of nucleation
5. composition factors enhancing the stability of the vitreous state
6. the effect of small quantities of additives on nucleation:
   (a) positive catalysis
   (b) negative catalysis.

Thus there are five methods available to reduce nucleation, that is to promote the initial process of crystallization, namely glass formation:

1. reduction of the liquidus temperature, that is of the temperature interval of crystallization
2. application of more complex compositions
3. reduction of the number of defects in the glass structure
4. application of LŐCSEI's "micro-eutectic" principle [108, 109]
5. enhanced masking of the cations with a tendency to phase separation, that is modification of the quality of cation and anion bonds, respectively.

On the basis of the nature of their nucleation effect the ions having a nucleation action in silicate melts may be classified into the following groups:

1. Fluoride nucleators

Under the action of heat treatment CaF_2 and NaF crystals will separate from the silicate melts. Rate of separation and the size of the crystals will depend on the composition of the base glass and on the fluoride concentration. In the systems SiO_2–Al_2O_3–CaO–MgO and SiO_2–Al_2O_3–CaO–Na_2O fluorides are highly effective nucleating agents. The nucleation action of fluorides may be considered as a process governed by structural defects (LUNGU, POPESCU–HAAS [49]).

2. Sulphide nucleators

CdS, MnS, FeS, CuS, CoS, MoS, etc. and the selenides belong in this group. The sulphide and selenide ions can be polarized to a much higher degree and can enter into far stronger interaction with ions having electron shells other than the rare gases than the oxygen ion. During the cooling process the energy difference between the $Cd-S$ and $Cd-O$ bonds exceeds the kinetic energy of the system whereby CdS or other heavy metal sulphide nuclei may be formed. An analogy to this process may be found in magma crystallization [1].

3. TiO_2 nucleators

Stable glasses are known in the system Na_2O–SiO_2–TiO_2, while from glasses with more complicated compositions rutile separates even at low, 4–5% TiO_2 content. At higher temperatures TiO_2 is soluble, at lower temperatures it separates from the melt, since the coordination of four formed at higher temperatures converts into a coordination of six at lower temperatures. The lattice defect structure of TiO_2 contributes to its nucleating action preventing the formation of large crystals in the course of TiO_2 separation, that is glass defects of "petrification" nature. A similar effect is also attributed to ZnO_2, SnO_2, MoO_3 and WO_3 (STOOKEY [27], HINZ [44]).

4. Metal nucleators

Gold, silver, copper, platinum, palladium, carbon, etc. may be classified in this group. These are in a colloidal dissolved state in the glass at high temperatures, while at lower temperatures they precipitate (STOOKEY [29]).

It should, however, be noted that the various nucleators display no identical actions independent of the compositions. In many compositions P_2O_5 acts as a nucleating agent, while it reduces the crystallization tendency of glasses with high alumina contents. Titanium dioxide may be considered a practically universal nucleator, but its effect is greatly reduced by the presence of sulphide, selenide and telluride ions. Accordingly, the presence of titanium dioxide reduces the nucleation effect in glasses nucleated with heavy metal sulphides. The nucleating effect of fluorides is greatly inhibited by the presence of boron trioxide [171].

Chapter V

Preparation of molten silicates

There is a certain difference, though not in principle, but in some technical details of the manufacturing technologies of molten rocks and vitroceramics. The manufacturing processes resemble mainly the technology of glass melting. The technology consists of the following part processes [1, 2, 3, 4]:

(A) preparation of the raw material
(B) melting
(C) processing of the melt
(D) molding
(E) crystallization
(F) cooling

(A) Preparation of the raw material

In the case of molten rocks raw material preparation consists in coarse comminution, in the case of synthetic raw materials the finely ground components must be weighed and homogenized in accordance with their chemical compositions accounting for the necessary stoichiometric proportions. This operation is exactly the same as the corresponding phase in glass manufacture. The raw rock is added in lumps.

(B) Melting

For the production of molten rocks at first glass furnace types were used, in France and Germany mainly the Siemens–Martin type furnaces. Much work has been done in the Soviet Union and Czechoslovakia to develop the construction of furnaces for molten rock production.

A number of constructional types have been tried and finally a shaft melting kiln was evolved which to some degree resembles the cupola furnaces used in iron production and in the preparation of slag wool. This furnace type was found the most advantageous from the aspect of energy consumption. It has a high melting rate and compared to its melting capacity the dimensions of the furnace are small.

In France and in the Soviet Union electrical kilns are also in operation. From the technological aspect these have the great advantage that due to a more rapid clarification they provide for an even higher melting rate [2, 150, 151].

In the Siemens–Martin type kilns the rock is added in lumps of 1–30 mm diameters, while in the shaft furnaces the charge is 70–150 mm large. For this reason melting in shaft furnaces is more economical, though comminuted charges preferably offer a possibility of any eventual correction of the composition and of the application of additives which is far more difficult in shaft melting where a homogeneous distribution of the additive is almost impossible [151].

The melting kilns operate, depending more or less on the basicity of the rocks, at temperatures between 1320 and 1400 °C. During the preparation of molten rocks the melting temperature must be observed within a relatively smaller interval (of 20–30 °C). It is not advisable to raise melting performance by raising the melting temperature, since higher melting temperatures may have an unfavorable influence on the subsequent crystallization process. With silica contents between 42 and 50% the melting temperature of the rock is linear between 1320 and 1400 °C, when a 42% silica content corresponds to a temperature of 1320 °C, and 48–50% of silica content to a temperature of 1400 °C.

The apparently most up-to-date rock melting in shaft furnaces utilize a significant proportion of the flue gases for the preheating of the material. The schematic diagram of such a shaft furnace is shown in Fig. 27.

Figure 28 shows schematically the construction of a higher capacity shaft furnace. Actual melting takes place in the opening at the bottom of the shaft and the melt flows from a homogenizing channel into a feed drum. In the shaft the material slides downwards at a rate corresponding to the speed of melting and the deficiency created in this way is made up by the continuous addition of more material on the top. For the production of molten rocks the kilns are lined with chrome magnesite since this material resists best the corrosive action of the alkaline melt. The adequate quality of the refractory is a constant problem in the production of molten rocks. It has been attempted to prolong the lifetime of the refractory by constructing the bridges and channels in such a way that the basalt melt shall solidify at the bottom and the melt shall flow over a solid rock bed protecting the bottom lining against corrosive influences.

One of the characteristics of rock melting is the incomplete melting of olivine and magnetite due partly to their high melting points and partly to the high rate of melting. This is a necessary feature from the aspect of processing, otherwise difficulties may arise in the formation of the crystalline structure. It has been observed that the melt used in processing is not completely homogeneous, but always contains crystalline nuclei. It has

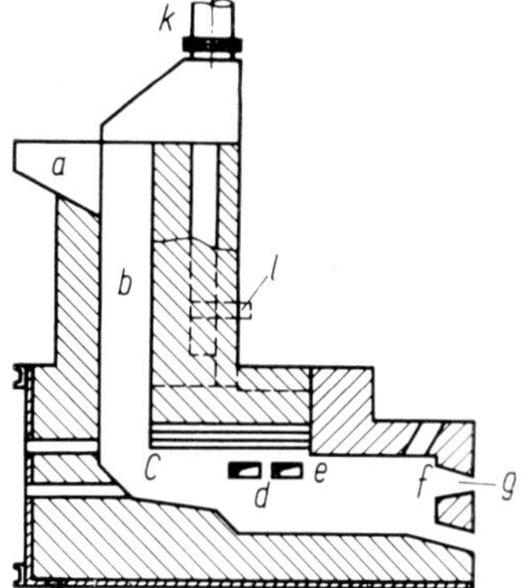

Fig. 27. Longitudinal cross-section of a Konarek type shaft furnace. a) Feeder, b) shaft, c) shaft opening, d) conditioning space, e) flue gas scum ports, f) and g) burner ports, k) flue off-sets, l) control slide valve

occasionally been attempted to improve the crystallization tendency by grinding the residual material at the bottom of the kiln when operation is interrupted. This residue is richer in olivine and magnetite because of a certain sedimentation and the ground residue is returned in small portions to the kiln or to the processing tank [151].

The refractory of the kiln which is in contact with the melt is usually chrome magnetite. More recently corundum and zirconium corundum have been successfully applied as refractories for this purpose. The resistance of the refractory is improved by building the kilns so deep that the basalt melt shall solidify at the bottom which saves the bottom lining from corrosive attack.

In addition to the recirculation of olivine-rich material additives are also used, or the raw material may consist of two types of rocks. However, because of the construction of the shaft furnace, these methods cannot be applied, since the coarse raw material lumps prevent the homogeneous distribution of the additive. In the case of shaft furnace construction the waste – corresponding to about 20–30% of the melted quantity, can neither be refed. This has several reasons; the vitreous basalt mellons rapidly and thus may cause a blocking of the shaft. Another technological difficulty arises from the short melting period which prevents the homogenization of the melt. Because of the second melting, the recirculation of the basalt will reduce the crystallization tendency of the material, which indicates a possible nucleation. Second melting reduces the quantity of olivine and magnetite which act as heterogeneous nucleating agents and this will result in crystallization hindrances [152–154].

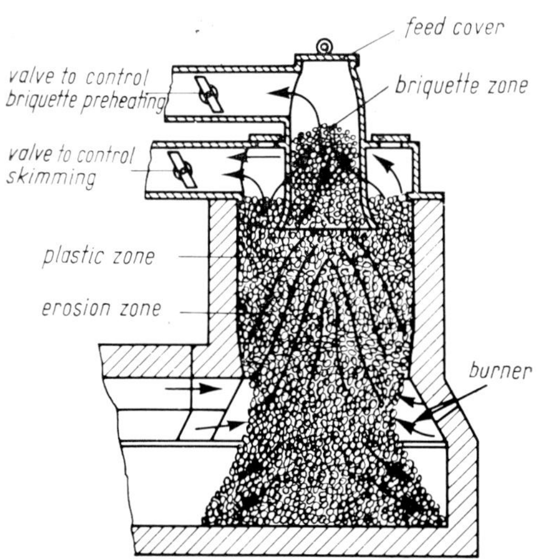

FIG. 28. Cross-section of a high capacity basalt melting shaft kiln

During the melting process if there is a possibility by technology, application of small quantities of additives was tried for several reasons:

1. Increase or decrease of rock basicity. In the Soviet Union for instance picrite (basic effusive rock) was added to diabase, in Czechoslovakia experiments were carried out with the combination of strongly basic tephrite and basalt [2].

2. Additives, e.g. soda or water glass, may be used to improve melting or processing conditions.

3. Additives will increase the quantity of small nuclei.

The melting temperature of the synthetic basic vitroceramic material on feldspar–diopside base is 1390–1400 °C. The melting conditions of such

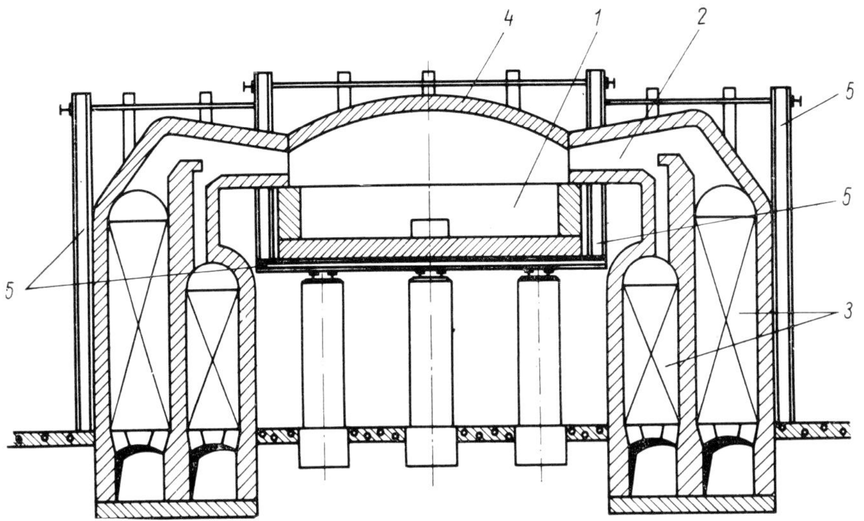

FIG. 29. Cross-section of a Siemens–Martin regenerative type tank furnace, 1. melting tank, 2. burner head, 3. regenerative chambers, 4. furnace roof, 5. steel structure of the furnace

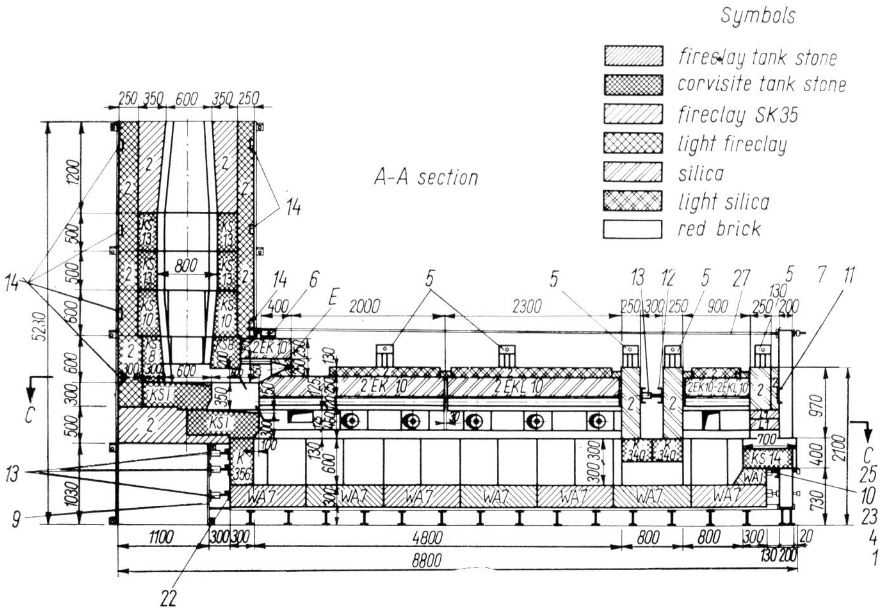

FIG. 30. Shaft furnace with conditioning and processing chambers

a synthetic material consisting of several raw materials are analogous to those encountered in glass technology. Melting may be performed in a Siemens–Martin regenerative furnace whose schematic diagram is shown in Fig. 29 [155].

Information may be gained by the conditions of melting technology of green glass for materials with sulphide nucleation, and by the melting conditions of brown and yellow glass for other vitroceramics. A shallow

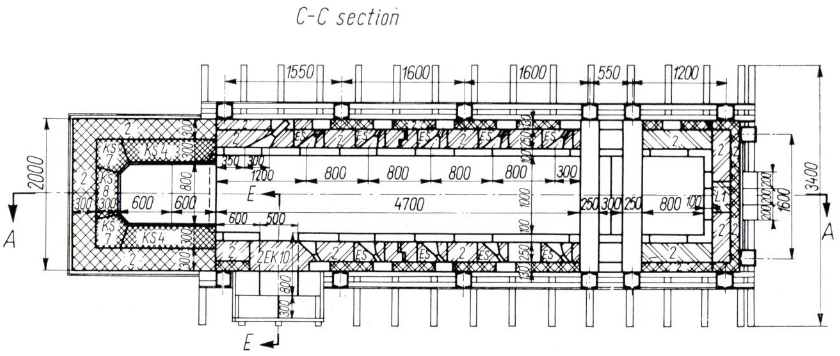

FIG. 31. Horizontal cross-section of a shaft furnace construction

tank is required, since owing to the iron sulphide or manganese sulphide content of the melt this will absorb high quantities of the radiation energy.

Shaft furnace construction may be applied to the melting of synthetic vitroceramic materials too, provided the appropriately comminuted mixture of the components is then briquetted. Shaft furnaces for the melting of synthetic vitroceramics are shown diagrammatically in Figs 30 and 31 [156, 157].

(C) Processing and shaping of the melt

In the case of vitroceramics the melt has to be processed at temperatures above the liquidus temperature contrary to the processing of rocks. Petrous materials are processed by casting. Molten rocks, slag and synthetic primarily crystallizing materials are cast into sand molds by a technology which is similar to that used in the iron and steel industries when at the same time the molding possibilities are also approximately analogous.

From the melting furnace the melt flows either into a tank or a horizontal drum. In the first case the melt is scooped into the molds, in the second it is poured. The drum can be rotated around its horizontal axis by means of a cogwheel transmitted tilting device and the melt can be poured from the same opening on the top of the drum which is also used as a melt inlet. The feeding drum is heated by separate burners. The most intricate phase in the production technology of molten rocks is the introduction of the melt into the molds. Up to now it was not possible to find a satisfactory mechanical solution by means of glass technological methods with the exception of readily melting vitroceramics nucleated only with metal and TiO_2. The temperature of the tapper drum is 1140–1150 °C. From the molten rock slabs and arched segments are cast. The molds are made mainly of sand and only seldom of metal, since the metal molds will

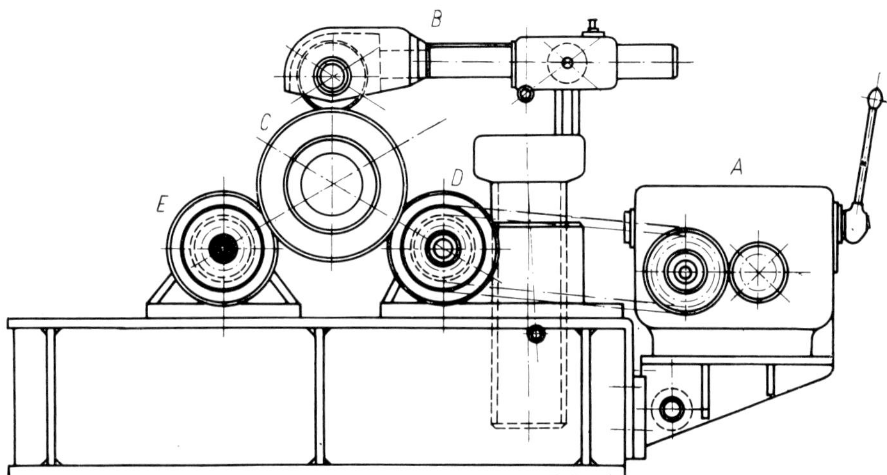

Fig. 32. Centrifugal pipe casting equipment, A) driving gear, B) clamping prop of centrifuge cylinder, C) centrifugal casting cylinder, D) driving roller, E) roller

considerably cool down the melt, causing disturbances in the course of crystallization. With the help of a centrifugal casting machine (Fig. 32) the molten rock can be processed into pipes. In France molten rocks have been processed in insulated and preheated metal forms by compression, this, however, requires special mold constructions and operations [3, 155].

Casting may also be applied to vitroceramics when the mold is made of metal. Because of the short useful interval of melt viscosity, molding possibilities are limited compared to the processing temperature of low viscosity petrous materials. For this reason vitroceramics are usually molded under pressure in metal molds. Blowing as applied in glass technology is applicable only to the so-called pyroceramics [158].

Molding of materials on feldspar–diopside base and prepared mainly by sulphide nucleation is illustrated on the following figures. Figure 33 is the picture of a valve body cast from a so-called crystalline synthetic stone (KM2) into a sand form together with the wood core. Figures 34, 35 and 36 are the auxiliary tools used in molding. Figure 37 is a sand blast nozzle cast into a sand mold, Figs 38, 39 and 40 the auxiliary tools used for the shaping of this nozzle.

4

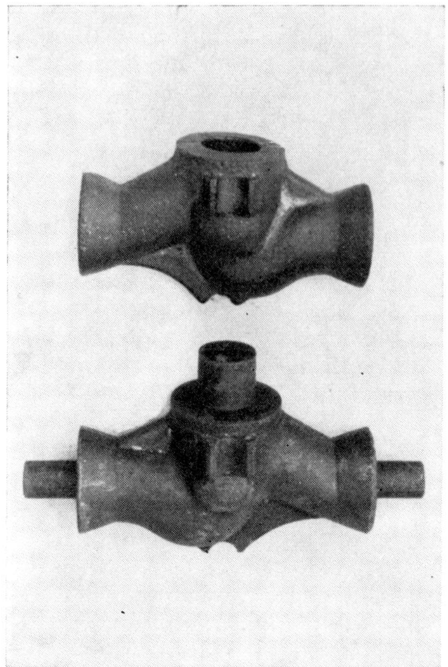

FIG. 33. Acid resistant valve body cast from the vitroceramic type KM3 and the internal wood core of the sand casting mold

FIG. 34. The prepared sand casting mold of the acid resistant valve body is seen in the two halves of the die

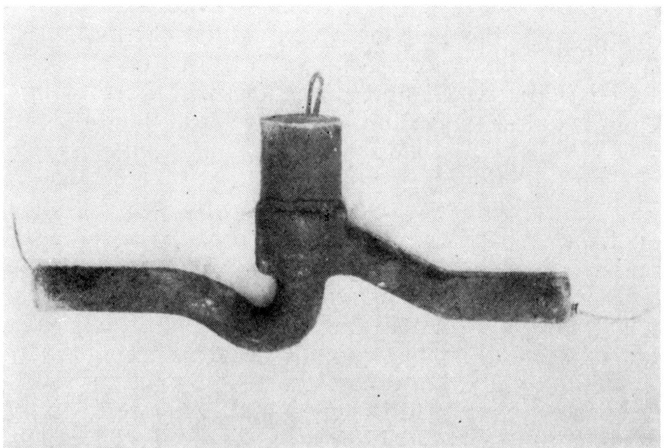

FIG. 35. Sand core to form the internal hole of the acid resistant valve body with the wire insert which helps molding

FIG. 36. One half of the die for casting acid resistant valve bodies and the internal core of the sand in the die

FIG. 37. Sand blasting nozzle made of material KM3

FIG. 38. The wooden auxiliary tool for the sand mold in which the sand blasting nozzle is cast

Fig. 39. Internal sand core and wood die for the casting of the sand

Fig. 40. One half of the die used for the casting of sand blast nozzles from material KM3 and the inserted internal sand cores

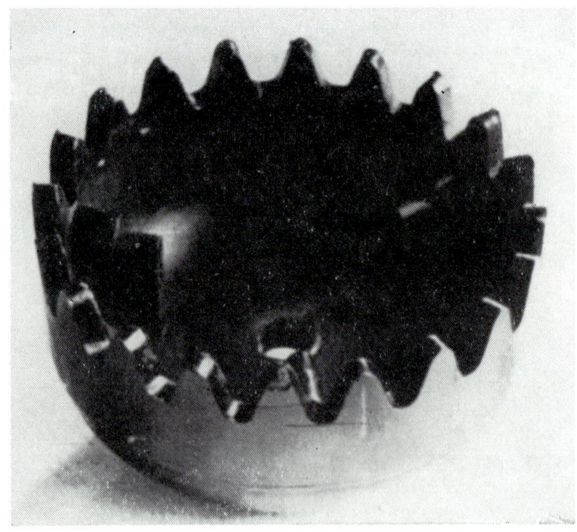

Fig. 41. Distillation cap pressure molded in a metal mold from vitroceramic KM2

FIG. 42. Electrical pile insulators made of vitroceramic KM3

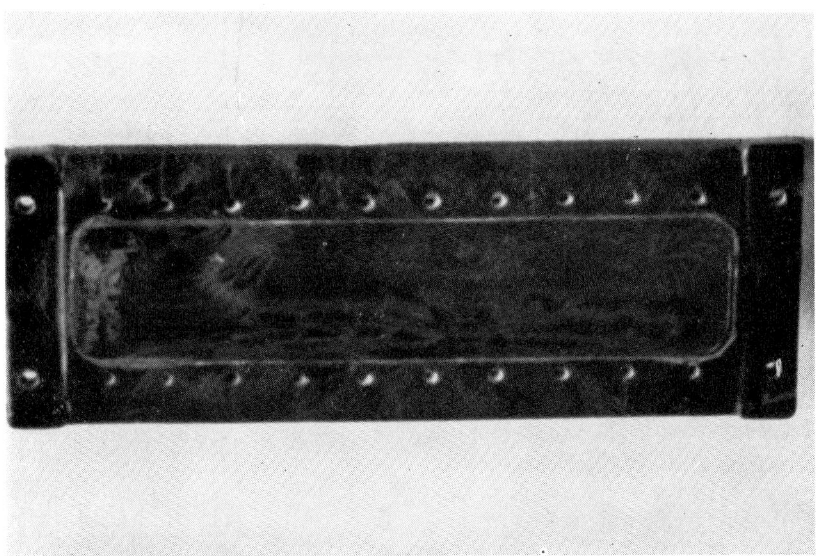

FIG. 43. Electrical cable holder pressure molded in a metal mold from vitroceramic KM2

Shaping possibilities are excellently illustrated by the vapour cap of a distillation column made of material KM2 (Fig. 41) which was shaped in a form mold under pressure. This shows that with an appropriately constructed die even intricate forms can be molded such as the toothed flange, the hole in the middle and the internal fins.

Figure 42 shows electrical pin insulators of various sizes molded under pressure in metal molds. Molding operation and the construction of the die have to account for the special properties of the material. A similar type of molding is shown in Figs 43 and 44. Figure 45 is the picture of a pressure molded gutter sink hole with a removable grate the whole made of material KM2 [1, 158]. The various other vitroceramic material types possess many further molding possibilities. Pipes and even cones may be made by centrifuging. Centrifugal casting is also applied to the shaping of pyroceramic types of materials. In this case the liquid, colorless glass flows into a conical steel form which rotates at a high speed around its vertical axis for a few seconds. The centrifugal force distributes the vitreous material rapidly and uniformly on the internal surface of the form and the material will cool rapidly. The vitroceramics used as rocket noses are shaped in this way.

For the molding of molten basalt metal molds may be used only after preheating to 450–550 °C. For the production of crystalline synthetic stone the mold has to be preheated to 200–300 °C, this temperature is provided by the heat loss of the melt fed in a continuous process.

FIG. 45. Gutter trap with inserted grate made of vitroceramic type KM2 by pressure molding. The surface of the grate is polished

In the Soviet Union sand molds are used for the casting of molten rocks. In France a vitreous residual layer is formed on the surface of basalt which has been cast into metal molds, being a technological necessity in basalt processing. The heat treatment of crystalline synthetic stone can be adjusted in such a way that no vitreous layer shall remain after crystallization on the surface of materials cast in metal forms. Casting into sand molds draws a parallel between the production of molten rocks and metallurgical technologies. The composition of crystallized synthetic stone allows to an increased degree the use of metal molds and shaping under pressure as applied in glass manufacture with the added advantage of greater possibilities of machine shaping.

The production technology of vitroceramics which have been nucleated with fluorides, metals or TiO_2 offers an even wider scope for the application of glass shaping methods, thus these materials may be blown and small diameter tubes may be produced from them by means of DANNER's or SHILLING's method.

From pyroceramic type base materials bottles, cylinders, dishes, rods, etc. may be prepared by blowing and even automatic glass shaping methods may be applied. By pressure shaping wide necked vessels, plates, tiles, crucibles, dishes may be obtained. Slabs, rods and fibres are prepared by extrusion [158].

(D) Crystallization

After shaping the structure of the molten silicate is formed by heat treatment. The bulk of the end product is crystalline with a residual amorphous phase whose quantity is determined by the nature of the composition and varies between 5 and 20%. Heat treatment has to provide partly for the completion of the crystallization process and partly for the gradual, stress-free annealing of the products. Crystallization of the petrous materials begins already during the shaping process, since the melt contains initially some nucleating agents and the crystallization tendency of the melt is high enough to provide for the primary commencement of crystallization.

In the production of molten rocks the product leaving the mold is in 60–70% of its mass already crystalline.

The composition of vitroceramics may be adjusted so as to avoid the formation of crystal nuclei during shaping, that is to prevent the beginning of the nucleation process. As a recent result it has been discovered that nucleation may begin during shaping with the advantage of a simpler and more rapid crystallizing heat treatment. The choice of the method depends on its expediency in the given case and on the application and size of the finished product. The heterogeneous crystallization process may be characterized by two parameters: by the number of crystallization centres per unit volume and by the rate of growth of the crystal nuclei. Depending on the composition of the shaped material heat treatment will begin between 500 and 750 °C and end between 700 and 1100 °C. The duration of heat treatment will depend on the wall thickness and linear dimensions of the product and may last from 5 to 10 hours, in case of special products even 20 hours. Heat treatment is followed by cooling of the crystalline materials, this process taking from 5 to 15 hours and only in exceptional cases more. For the annealing process preferably continuously operating tunnel furnaces are used. Low tunnels are the most appropriate and necessary. "Sandwich" heat treatment [159] has often many advantages, since it provides best for the uniform heating of the material in the furnace [154–158].

The processability of rocks and slags is, however, determined by a number of factors, consequently several materials melted independently of their composition will not lead to the desired product.

In the crystallization of vitroceramics the composition has to ensure the absence of deformation of the molded product during heat treatment. Deformation depends on the correlation between viscosity and temperature and on crystallization tendency and nucleation efficiency. At a viscosity value of $10^{7.6}$ poise the own weight of the material will cause its deformation. When loaded with 10 kp/mm² the glass fibre will display an elongation of 1.0–1.5 cm per minute, at $10^{8.6}$ poise the rate of deformation is only one-tenth of the above. Materials with 10^{9-10} poise viscosity will show a deformation only after prolonged application of great loads. Thus for deformation-free crystallization the viscosity of the petrous materials must reach a value over 10^{10} as a result of crystallization at the initial temperature of heat treatment, that is at 700–800 °C. In the annealing furnace the vitroceramics may be heated to 10^{9}–10^{10} poise viscosity when the crystallization rate of the glass and the separation of the nucleator have to provide for the viscosity increase, since the temperature has to be raised in order to reach the favorable rate conditions for crystallization. If the viscosity increases during heat treatment, crystallization rate will drop proportionally, since the diffusion rate required for crystallization is inversely proportional to the viscosity. Temperature has to be raised on the one hand up to a point when nucleus formation and the growth of crystal nuclei proceed at the desired rate, while on the other hand the viscosity of the glass must always exceed 10^{9-10} poise.

The criterion for the formation of vitroceramics is the beginning of crystallization within the glass. Crystallization starting from the surface may have two reasons:

1. lack of nucleating agent, that is its insufficient quantity or inadequate quality;
2. a too high crystallization tendency of the glass.

Hence the crystallization tendencies of petrous and vitroceramic materials have to meet different requirements.

Petrous materials have high crystallization tendencies, where the high crystallization rate dominates, though the presence of a certain amount of crystal nuclei is indispensable. In the formation of vitroceramics a lower crystallization tendency is permissible only, since the formation of a highly disperse structure is bound to the condition that the crystallization of the material shall not begin during the shaping process; the material has to solidify first in the vitreous state, while its nucleation tendency must be absolutely greater than the crystallization tendency of glass. In this respect nucleation is of particular importance which has to be much higher than in the case of glass; this nucleation is ensured by the addition of nucleating agents.

Information was gained on the crystallization conditions of molten silicates among others by an experiment carried out in a temperature gradient furnace in which the beginning of crystallization and the changes in crystallization rate were observed. For the experiment a glass rod which had been heat-treated in a platinum trough was used and crystallization was observed by means of the microscope.

Information on the crystallization and viscosity changes was further obtained by the LITTLETON type softening point determination of a glass filament drawn from the melt. The beginning of elongation corresponded to $10^{8.6}$ poise viscosity. At the corresponding temperature a 26.8 mm long filament suffers in a vertical tube furnace a 1-cm elongation under its own weight in 1 minute. Softening points are quite often observed for materials in the lithium–aluminum–silicate system which had been nucleated with

silver, copper, gold or platinum. The experimental results are shown in
Fig. 46.

Generally no softening points are found for materials on furnace slag or
rock base and nucleated with sulphide or fluoride. The results of the examination of their softening points are shown in Fig. 47. There are certain

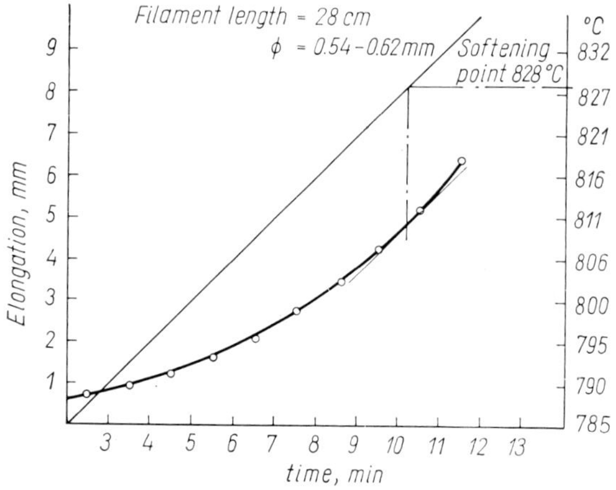

FIG. 46. Determination of the softening point of
vitroceramic base glass

experimental results indicating a constriction instead of an elongation of
the filament due to contractive forces during crystallization.

Still further information may be gained on crystallization conditions,
on the nature of nucleation and on temperature parameters by means of

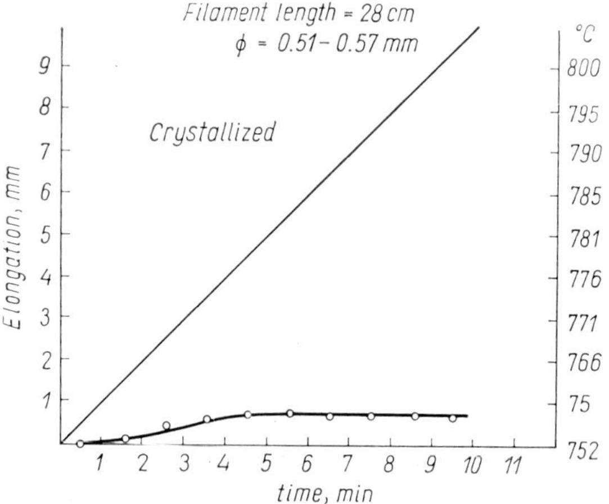

FIG. 47. Changes in the softening of a vitroceramic
glass on furnace slag base and nucleated with
sulphide

DTA (differential thermoanalytical) tests, as illustrated by the DTA curves of the base glass of a synthetic molten crystallized silicate on slag base (synthetic stone) and of a lithium aluminum silicate glass showing the beginning of crystallization in the two systems (Figs 48 and 49).

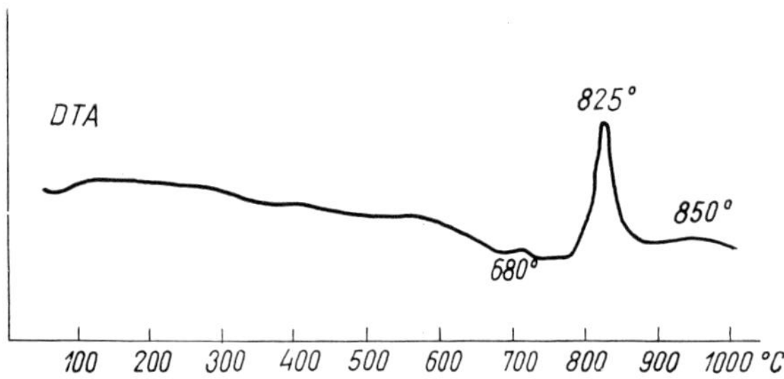

FIG. 48. DTA diagram of a vitroceramic glass on slag-rock base

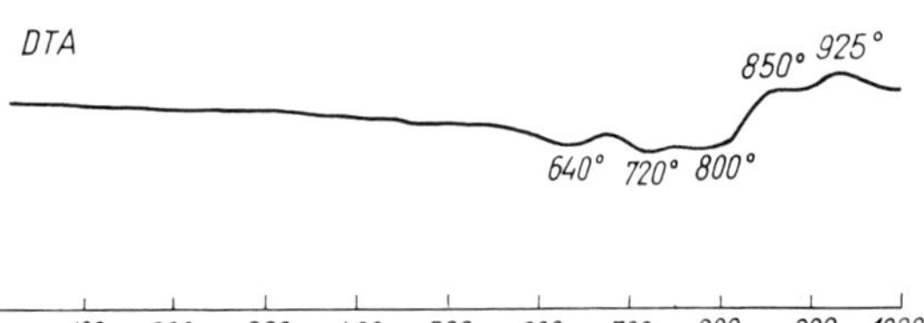

FIG. 49. DTA diagram of a lithium aluminum silicate vitroceramic base glass

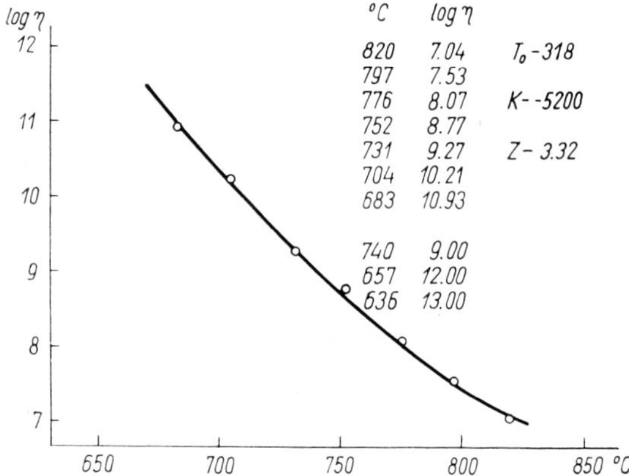

FIG. 50. The viscosity of the vitroceramic base glass vs. the temperature in the softening range of the glass

Changes in viscosity in the temperature range under investigation are further important data. Viscosity vs. the temperature was plotted in the viscosity interval from 10^{12} to 10^7 poise as illustrated in Fig. 50 for the vitroceramic glass of the lithium aluminum silicate type. From this figure and from the DTA curve in Fig. 49 the following conclusions can be drawn with respect to the heat treatment program: up to the temperature pertaining to $10^{9.5}$–10^{11} poise heating rate may be high, the material may be kept without deformation for some time at lower temperature, but in this

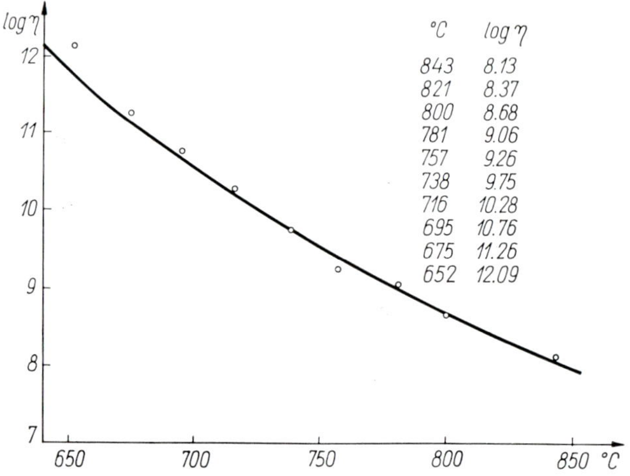

FIG. 51. The viscosity curve of the lithium aluminum silicate glass in the softening range

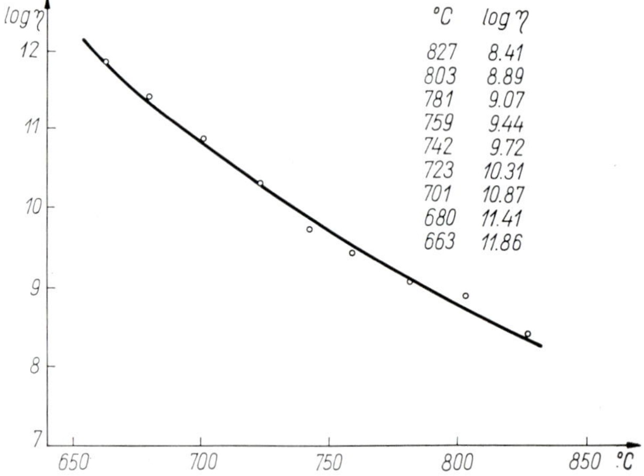

FIG. 52. The viscosity of the vitroceramic glass on lithium alumium silicate base after 30-minute preliminary heat treatment at 650 °C

viscosity range crystallization will proceed but slowly. Heat treatment will induce crystallization and a consequent rise in viscosity, and the deformation limit will gradually shift towards higher temperatures. This is illustrated by the following example: the viscosity of the glass was determined after heat treatment in the temperature range of softening. Figure 51 shows the softening viscosity curve of a glass in the lithium–aluminum–silicate system without preliminary heat treatment, while Fig. 52 is the softening viscosity curve of the same glass after heat treatment at 650 °C for 30 minutes. The

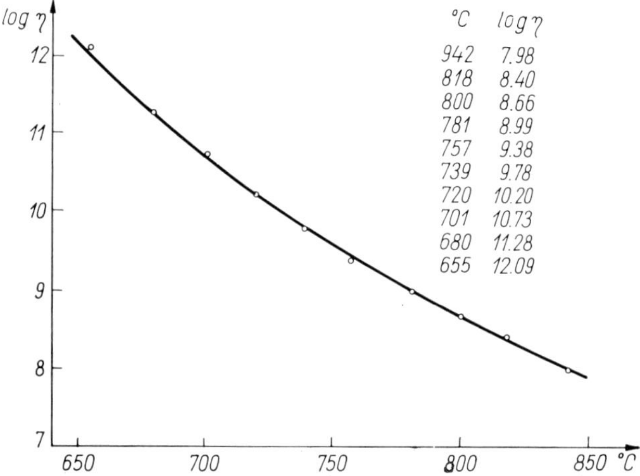

Fig. 53. The viscosity of the vitroceramic glass on lithium aluminum silicate base after preliminary heat treatment at 650 °C for 90 minutes

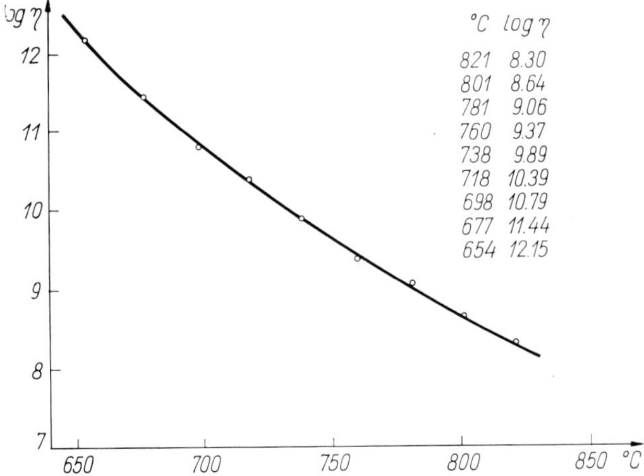

Fig. 54. Viscosity of the lithium aluminum silicate base vitroceramic glass after 30-minute preliminary heat treatment at 720 °C

temperature values pertaining to 10^{13}, 10^{12} and 10^9 poise, respectively, were chosen as the main characteristic values of the viscosity curve. The temperature of 650 °C corresponds to a viscosity of 10^{12} poise, that is to the lower limit of the nucleation range.

The viscosity equation is:

$$\eta = k\,\mathrm{Exp}\,\frac{Z}{T - T_0} = -\,3.9 \times 10^4\,\mathrm{Exp}\,\frac{1.01}{T - 300}$$

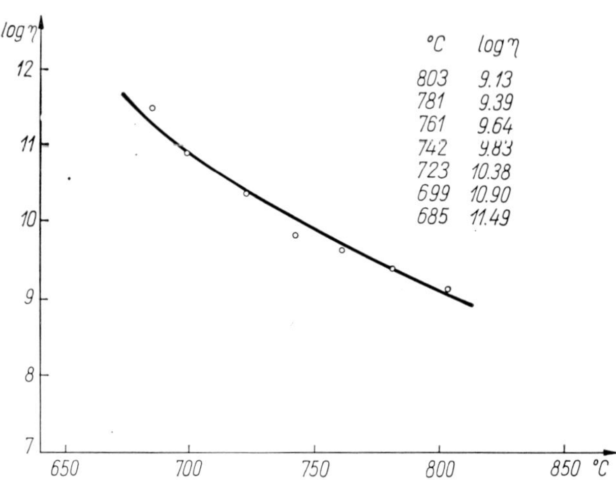

FIG. 55. Viscosity of the lithium aluminum silicate base vitroceramic glass after 30-minute heat treatment at 760 °C

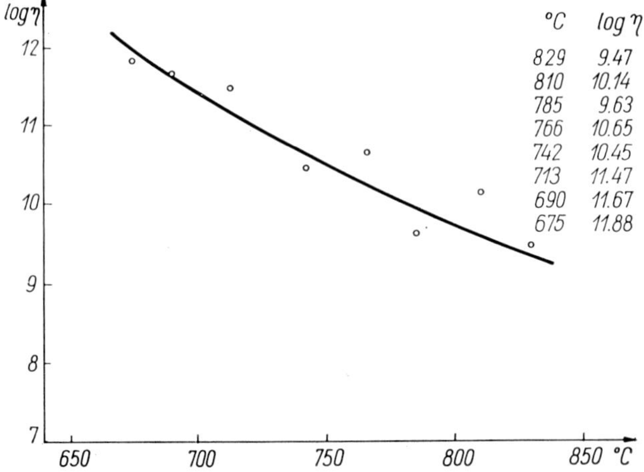

FIG. 56. Viscosity of the vitroceramic glass on lithium aluminum silicate base after 60-minute preliminary heat treatment at 760 °C

where k is a constant; η is the viscosity; T is the temperature in °K; T_0 is a constant and Z is a value proportional to the activation energy.

Figure 53 shows the viscosity curve after a 90-minute heat treatment at 650 °C for which the equation of the viscosity is:

$$\eta = -\ 3.7 \times 10^4 \times \mathrm{Exp}\ \frac{1.04}{T - 313}\ .$$

The temperatures pertaining to the chosen fixed points agree within the limits of the experimental error.

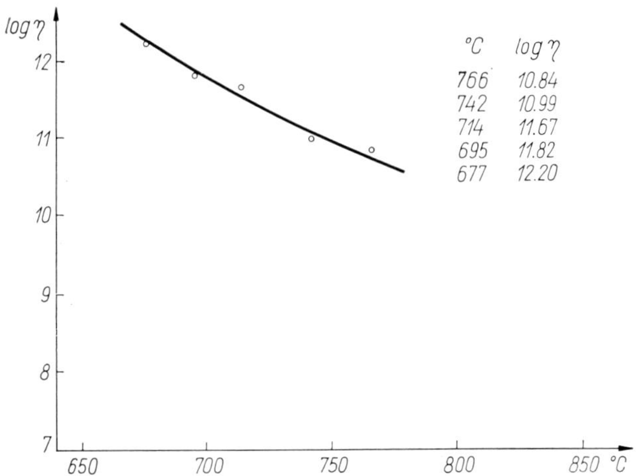

Fig. 57. Viscosity of the vitroceramic glass on lithium aluminum silicate base after 90-minute preliminary heat treatment at 860 °C

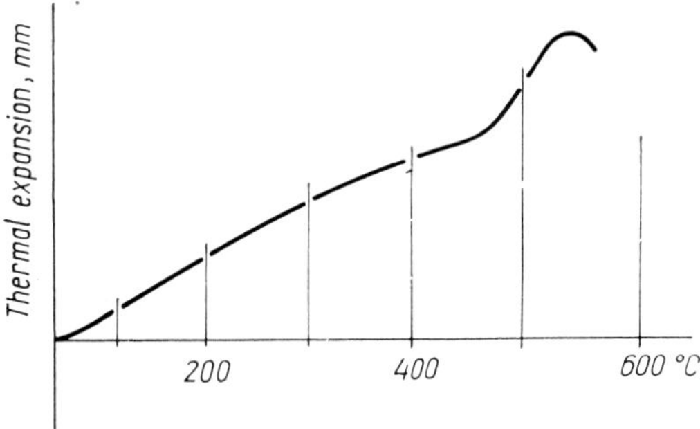

Fig. 58. Heat expansion curve of the vitroceramic base glass

The viscosity curve in Fig. 54 was obtained after 30-minute heat treatment at 720 °C for which the viscosity equation is

$$\eta = -\,4 \times 10^4 \times \mathrm{Exp}\,\frac{-\,0.4}{T - 310}.$$

The temperatures pertaining to the fixed points are unchanged, thus heat treatment caused no change in viscosity.

Figure 55 illustrates the conditions after 30-minute heat treatment at 760 °C, and Fig. 56 after 60-minute heat treatment at the same temperature. Here already a considerable change may be observed indicating the beginning of crystallization. The viscosities of the chosen points rise significantly after 30- and 60-minute heat treatments. Figure 57 illustrates the softening viscosity of glass after 90-minute heat treatment at 860 °C. The viscosity equation is

$$\eta = -\,7.0 \times 10^4 \times \mathrm{Exp}\,\frac{-\,0.59}{T - 50}.$$

The temperatures pertaining to the fixed points, mainly that pertaining to 10^9 poise, has risen further. Consequently it is possible to perform heat treatment without deformation even at 900 °C, and the heat treatment program of crystallization can be as follows:

Up to 660–700 °C heating may be rapid. This is the nucleation range and the glass should be kept in this temperature interval for 30–40 minutes, followed by a heat treatment at 700–760 °C for 60–90 minutes. The temperature is then raised to 860 °C in 60 minutes and further heat treatment is to be determined in accordance with the special requirements of crystallization.

Hence the general program of crystallization for vitroceramics is as follows:

I. Heat treatment in the nucleation range: 30–60 minutes, temperature effect pertaining to 10^{12}–10^{11} poise.

II. Heat treatment to induce crystallization in the viscosity range of 10^{11}–$10^{8.5}$ poise, duration of heat treatment: 60–90 minutes.

III. Heat treatment in the 10^8 poise viscosity range of the base glass.

IV. Heat treatment at a temperature corresponding to the 10^8–10^7 viscosity interval of the base glass.

V. Gradual cooling which may include a shorter or longer heat treatment at the temperature corresponding to 10^8 poise viscosity, or the reduction of the cooling rate to a minimum value.

Figure 58 shows the results of the dilatometric measurements of a vitroceramic base glass vs. the temperature. The dilatometric tests indicate the dilatometric softening point whose viscosity value is 10^{11}–10^{12} poise, thus corresponds to the nucleation temperature range of the materials. In the given example the temperature is 572 °C, consequently the nucleation temperature range may be primarily determined as 560–600 °C.

The effect of nucleation is shown beside the viscosity tests also by differential thermal analysis. Figure 59 shows the results of the differential thermal analysis of a lithium aluminum silicate glass cooled to the vitreous state, while Fig. 60 is the DTA curve of the same glass nucleated with TiO_2 and Ag after heat treatment at 600–650 °C. In this second curve the endothermic and exothermic effects appearing in the given temperature range on the first curve are no longer visible which may be interpreted as

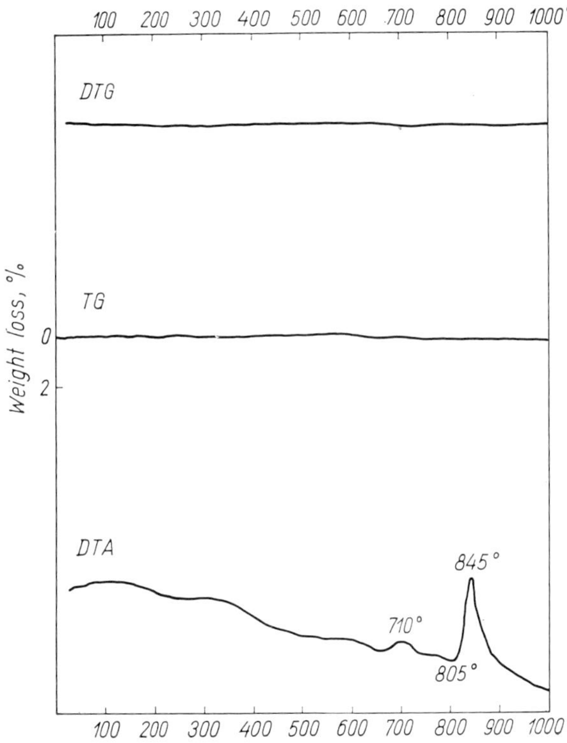

FIG. 59. The DTA curve of the vitroceramic glass on lithium aluminum silicate base

the activation and separation of the nucleating agent, indicating that nucleation took place under the effect of heat treatment in the temperature range between 600 and 650 °C. The effect of nucleation is also manifest in other physical properties of the glass. The light transmittance of the glass (SiO_2 68%, Al_2O_3 17%, TiO_2 5%, MgO 4.5%, $+ LiO_2 + Na_2O$ 4.5%) in the 500–600 μm range drops from 75% to 28%, while the density of the glass rises from 2.40 g/cm^3 to 2.46 g/cm^3. The heat expansion coefficient drops from 42.3×10^{-7} mm/mm °C to 41.27×10^{-7} mm/mm °C. The crystallization process may be followed from the changes in the heat expansion coefficient and the density and from the determination of these data the appropriate level of the crystallization process may also be checked. In the above example at the end of the crystallization process the heat expansion coefficient was 19.5×10^{-7} mm/mm °C and the density 2.51 g/cm³ [171].

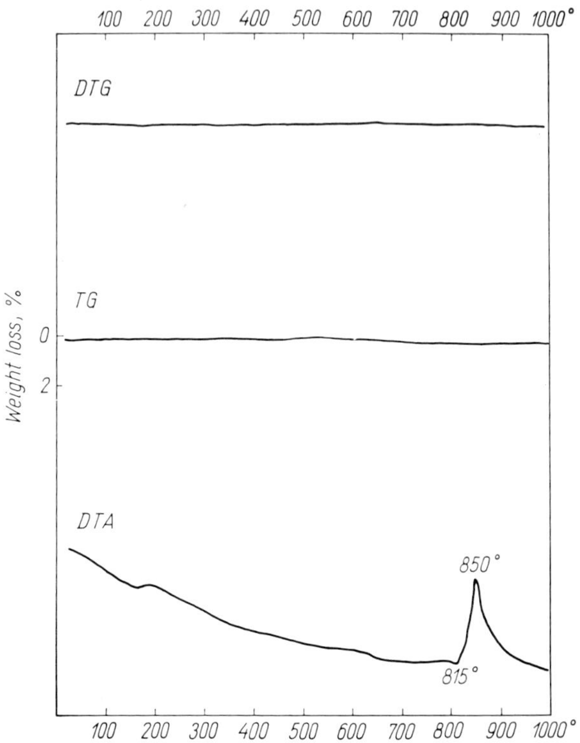

FIG. 60. The DTA curve of the vitroceramic glass on lithium aluminum silicate base after heat treatment at 600–650 °C

From these results we may sum up the test methods which may be applied if necessary on a plant scale to the detection and control of the nucleation and crystallization processes:

1. Plotting of the heat expansion curve of the base glass up to the dilatometric deformation temperature.
2. Differential thermoanalysis to determine the enthalpy changes in the glass, measurement of softening viscosity without preliminary heat treatment and after heat treatments at different temperatures, determination of the specific weight and heat expansion coefficient before and after heat treatment, that is in the various stages of heat treatment.

Chapter VI

Composition of vitroceramics
on feldspar-diopside base

(A) Deduction of the composition of crystallized synthetic stone

The results of acid resistance and mechanical strength tests of basalts have conveyed the suggestion that to work out molten silicates with improved properties compared to basalt in order to widen the field of application of these materials. Synthetic preparation seemed to be soluble on furnace slag basis. Modification of the composition and changes in the crystallization mechanism opened ways to the improvement of chemical resistance and mechanical properties. Modification of the chemical composition involved first of all an increase in silica content, while the improvement of mechanical properties demanded finer structure, that is the decrease of the crystal sizes constituting the material.

In the composition of basalts iron oxide is the component reducing chemical resistance, as indicated by the chemical resistance tests of the individual mineral components of basalts. Mineralogically the basalts are composed of silicates and non-silicates. Into the first group belong the plagioclases, pyroxenes and olivine, into the second group magnetite and ilmenite. Thus iron is present partly in its oxide form and partly bound to silicate in the pyroxenes. Magnetite is readily soluble, ilmenite somewhat more poorly soluble in hydrochloric acid. Both constituents have a poor acid resistance, while the acid resistance of pyroxenes is far more favorable, but olivine is also the more readily soluble in hydrochloric acid the more fayalite it contains. Comparison of the acid resistances of bronzite and hyperstene reveals some interesting facts. Bronzite is not attacked by acids, while hyperstene is slightly soluble in hydrochloric acid. The iron oxide content of bronzite is 15–30%. These facts indicate that iron oxide content even when bound to silicate will greatly impair the acid resistance of the molten rock. The role of the nature of alkali metal oxides on chemical resistance may be demonstrated on hand of the different behavior of nepheline and leucite. The first is gelatined by concentrated hydrochloric acid, the second is dissolved without forming gel. Thus the alkali metal oxide is more favorable when it is a sodium oxide, than in the form of potassium oxide.

The aforesaid may be considered as a basis for the determination of the synthetic composition. The actual composition was, however, later influenced by certain technological necessities, such as fusibility, formability and production costs. From this aspect furnace slag being a waste product may be considered favorable. It has a low iron oxide content not exceeding 1–2%. The higher silica content is also an advantage from the point of view of chemical resistance and is only limited by fusibility. Composition is however also influenced by the crystallization tendency of the material and a balance between the silica content and the crystallization conditions has to be found.

The transformation of the macroscopic heterogeneity of molten rocks into a microcrystalline structure may be realized by reducing the crystallization rate and increasing the quantity of crystal nuclei per unit volume. Reduced crystallization tendency may be achieved by raising the ratio of network forming and modifying ions. The crystallization conditions may further be adjusted by varying the ratio of silica to alumina within the network forming constituents.

Finer molten silicate structures may be realized through the analysis of the crystallization tendency as first defined by TAMMANN, that is by the modification of the crystallization mechanism. The molten silicates prepared earlier crystallized primarily, that is crystallization occurred in 60–100% during the cooling process. If crystallization rate and changes in the number of crystal nuclei are studied from Fig. 6, we find that crystallizations of this nature are usually characterized by a relatively high crystallization rate and a small number of crystal nuclei. The dispersity of the structure may be enhanced when crystallization proceeds with a great number of crystal nuclei and at a lower rate. This may be realized if the crystallization of the material does not begin during the cooling and the crystallization region (according to Fig. 6 in the temperature interval $T_L - T_A$) is approached from the T_A direction. In other words: the material is first cooled to the vitreous state and then heated by an appropriate treatment to the temperature interval in which the desired structure will develop. In this way the material may be crystallized in the temperature range corresponding to the maximum number of crystallization nuclei. This operation, however, cannot be performed in the case of normal glasses without involving a certain deformation.

A component is needed which will separate primarily from the mass and will as a new phase fulfil the role of a nucleating agent, that is, the primarily separated crystals will in the following be evenly distributed in the entire mass of the material where they will form highly dispersed heterogeneous crystallization nuclei thereby accelerating and in fact determining the crystallization process.

In furnace slag this role was fulfilled by the iron and manganese sulphide content and the finely dispersed carbon. In spite of its small quantity the Cr_2O_3 content of the slag must not be left out of the considerations, similarly to the dissolution of the chrome magnesite refractory.

Table 4 shows the average compositions of the furnace slags from the Diósgyőr and Ózd Metallurgical Works. Furnace slag alone does not meet the requirements of material composition as mentioned above, and is therefore not suited for processing to secondarily crystallizing molten silicate, that is to vitroceramic material. It has but a low chemical resistance and the ratio of the network forming ions to the modifying ions is low. Furnace slag will crystallize primarily and lead to a coarsely dispersed structure. If the calcium oxide content of the slag exceeds 43%, calcium oxide may separate as a result of calcium silicate decomposition [161–163]. It has the further disadvantage that due to the transformation of the dicalcium silicate

TABLE 4

Composition of Diósgyőr blast furnace slag and of Hungarian rocks

Components	1	2	3	4	5		
	%				A_k	b	b_k
SiO_2	45.87	57.50	46.05	48.58	33.67	±2.32	0.67
Al_2O_3	15.20	20.50	17.92	12.70	11.37	±2.09	0.60
Fe_2O_3			4.03	11.61			
FeO	10.06	2.40	5.22	1.31	0.94	±0.71	0.20
MgO	8.29	0.30	6.51	5.41	5.82	±1.13	0.33
CaO	9.21	1.90	9.20	9.46	37.38	±2.11	0.61
Na_2O	4.29	8.80	5.32	3.44	—	—	—
K_2O	2.13	5.30	2.35	0.14	—	—	—
MnO	—	—	—	—	4.21	±2.84	0.82
TiO_2	0.55	0.40	1.82	2.40	0.35	±0.18	0.06
P_2O_5	0.08	0.02	0.05	0.35	0.60	±0.20	0.05
Cr_2O_3	—	—	—	—	0.08	±0.04	0.01
BaO	—	—	—	—	2.27	±0.92	0.27
S	—	—	—	—	1.50	±0.15	0.04
F	—	—	—	—	0.30	±0.19	0.06
SO_3	—	—	—	—	1.10	±0.81	0.23
Ignition loss	—	—	—	—	−2.80	±1.92	0.55

Arithmetical mean value A_k

Experimental scattering: $b = \pm \sqrt{\sum_{i=1}^{4} (A_i - A_k)^2 \cdot \frac{1}{n-1}}$

Average error of mean value: $b_k = \dfrac{b}{\sqrt{n}}$

1. Badacsony basalt
2. Hird phonolite
3. Nógrád (Medve) basalt
4. Szarvaskő diabase
5. Diósgyőr furnace slag

which is formed in the course of crystallization the shaped product may crack during cooling [8].

Furnace slag is used in the cement industry in a finely ground state because of its hydraulic binding ability. It is easily understandable that this material will not meet the requirements of molten silicate structural materials.

The composition of the furnace slag was modified by the addition of sand and soda, that is of Na_2SO_4 which has been reduced with carbon. Two types were worked out, the first type had the same expansion coefficient as iron [79], the second was a high abrasion and acid resistant material.

The compositions of these modified furnace slags are shown in Table 5. These materials were named crystallized synthetic stones [21, 33, 80, 81]. In the case of type KM1 modification of the composition was manifest first of all in a higher alkali content, since the primary aim was to produce a material which has almost the same heat expansion coefficient as iron, namely 110–120×10^{-7} mm/mm °C. In the second case (material KM2) the silica content was raised to 56–60%. The material contained of the alkali metal oxides either exclusively or mainly sodium oxide because of the solubilities of the various alkali silicates. The composition included sodium sulphide and rocks with alkali metal oxide content. It is one of the important features of the composition that the total quantity of silica and alumina must be between 62 and 65% in order to obtain deformation-free crystallization. According to Lőcsei's investigations [1, 100, 163] in the KM2 type

TABLE 5

Various "crystallized synthetic stone" compositions and composition limits in per cent

Components	KM1	KM2	KM3	KM1	KM2	KM3
SiO_2	30 —34	55 —58	53 —62	32.7	56.2	60.9
Al_2O_3	7 —10	6.5— 9.5	6 —16	8.6	7.9	13.2
MnO	3 — 5	2 — 4	2 — 5	5.5	2.9	2.8
FeO	0.5— 2	0.5— 2	0.2— 1.5	0.4	1.9	2.5
CaO	30 —37	20 —28	5 —20	37.0	22.6	9.2
MgO	3 — 8	2 — 6	5 —25	3.3	3.1	5.5
BaO	0.5— 2	1.5— 3	0.5— 3	0.4	1.7	0.3
K_2O	0.2— 1.0	0.1— 2	0.1— 3	0.2	0.7	1.8
Na_2O	10 —12	4.0— 7.0	1.5— 5	10.8	4.9	2.8
TiO_2	0.2— 1.0	0.2— 1.0	0.2— 1.0	0.4	0.6	0.5
Cr_2O_3	0.1— 0.3	0.1— 0.3	0.1— 0.3	0.1	0.1	0.1
P_2O_5	0.1— 1.0	0.1— 1.0	0.1— 1.0	0.5	0.4	0.3
F	0.1— 1.0	0.1— 1.0	0.1— 3.0	0.3	0.1	0.8

TABLE 6

Crystalline phases detected by Debye–Scherrer diagrams at the beginning of the crystallization of crystallized synthetic stone (KM2)

dkX	Intensity	Material	Heat treatment (minutes)
2.985	2	FeS	
2.687	1	MnS, FeS	1
2.469	2	FeS	
2.759	2	FeS, FeS_2	
2.456	2	FeS	
1.981	1	FeS	2
1.869	1	MnS, FeS_2	
3.054	1	MnS	
2.776	1	FeS, FeS_2	
2.059	3	FeS	3
1.938	2	MnS	

material which has first solidified in the vitreous state under the influence of heat treatment at 760 °C a new phase of uniform distribution will separate. The DEBYE–SCHERRER diagrams of the new phase reveal its crystalline nature. The lines of the diagram are given in Table 6. Further heat treatment at high temperature ensures the homogeneous microcrystalline structure of the material. The quantity of FeO and MnO, that is of iron and manganese sulphide necessary for nucleation was determined experimentally. The experimental results are summed up in Table 7, where the following symbols were used to denote the quality of crystallization: F is surface crystallization, a typical sign of the lack of nucleation, or of the unsatisfactory or insufficient nature of nucleation. This was found in the case of samples Nos 1, 2, 3, 4 and 10.

G stands for a poor crystal separation in the entire volume of the mass (homogeneous nucleation) accompanied by surface nucleation, as observed in the case of sample No. 5. From the comparison of this to the crystallization nature of sample No. 4 it appears that manganese sulphide is a more efficient nucleating agent than iron sulphide. Comparison of the behaviors of samples Nos 11 and 12 reveals also the nature of nucleation when pure nucleation may be explained by the low sulphide content of the glass. K denotes a homogeneous microcrystalline structure as observed in samples Nos 6 and 8, while J stands for a highly favorable crystal structure. In the case of E crystallization occurs primarily, and crystallization tendency is higher than required.

TABLE 7
Nucleating effect of iron and manganese sulphide

Oxides	Number of experiment													
	1	2	3	4	5	6	7	8	9	10	11	12	13	14
SiO_2	$56.5 \pm 1.5\%$													
Al_2O_3	$7.0 \pm 0.5\%$													
CaO	$25.0 \pm 1.0\%$													
MgO	$6.0 \pm 0.5\%$													
Na_2O	$5.5 \pm 0.5\%$													
S^{2-}	0.5	1.0	2.0	1.0	1.0	2.0	2.0	1.5	1.5	0.2	0.2	2.0	2.0	2.0
FeO	0.2	0.2	0.2	1.0	0.2	1.0	1.0	2.0	1.0	2.0	0.2	0.2	1.0	1.0
MnO	0.0	0.0	0.0	0.0	1.0	1.0	2.0	1.0	3.0	0.0	2.0	2.0	3.0	6.0
Nature of crystallization	F	F	F	F	G	K	J	K	J	F	G	G	J	E

But for the already mentioned fluctuations the oxide compositions of the experimentary materials were identical. Deviations may be ascribed to the fluctuating compositions of phonolite and feldspar and to variations in the production conditions of the melt. Manganese was introduced in the form of pyrolusite and the quantity of iron oxide adjusted by varying the feldspar : phonolite ratio. This series of experiments clarified the basic principles underlying the mechanism of the crystallization process. Detection of the nucleating effect of iron and manganese sulphide made it possible to modify the composition of this type of vitroceramics between wider limits and to produce them independently of blast furnace slag [1, 20, 57, 163].

(B) Comparison between the compositions of crystallized synthetic stone and molten rocks

As a basis of comparison the anorthite–allite–diopside system was used which had been investigated by BOWEN [165–168] and first suggested by ORMONT [9, 10]. Because of its complex nature no realistic conclusions can be drawn from the oxide system which in itself is suitable for individual comparisons only. From the comparison of oxide compositions no quantitative data can be determined with respect to certain particular properties of the material, but alone the direction of the change of certain properties can be predicted. Since in the comparison of different compositions several changes of identical nature or of opposite effect must be accounted for, it is rather difficult to determine the characteristic change of the final resultant.

More information may be gained from comparison on the basis of mineral composition. As already pointed out in the detailed discussion, the SiO_2, Al_2O_3, CaO and MgO content is contained in three minerals which constitute about 80–90% of the rock. Ferrous and ferric oxide and the small quantity of titanium dioxide are in the main the components of magnetite and ilmenite, and only to a small extent of titanium augite, olivine and other pyroxenes. The total quantity of metal oxides occurring in addition to the above is not more than 1%.

The evaluation of the albite–anorthite–dipside system is rather cumbersome, since each composition has to be converted into the mineralogical composition, but ORMONT's procedure considerably simplifies the operation [10, 11]. On the GIBBS triangle the apices correspond to the pure substances Ab, An and Di, their quantity decreasing from 100% at the respective apex along the sides of the triangle to 0% at the other two apices. If the apices represent the chemical composition of anorthite, albite and diopside respectively, that is the various quantities of the five oxides diverging from 100, the diagram will be symmetrical. If the compounds contain two common oxide components each, as in the given case, the detailed phase diagram of the five-component system may be presented in the form of a triangular diagram.

This principle may be applied to the study of molten rock compositions by reducing the compositions of the rocks to an iron oxide free form, whereby the location of the rock may be found in the albite–anorthite–diopside three-component system.

The reduction of the composition differs from the method applied by ORMONT, since the total FeO and Fe_2O_3 content has not to be subtracted from the chemical composition, but only a quantity present according to its stoichiometric ratio in magnetite and by accounting for the titanium dioxide content of ilmenite. According to LEONTEEV's data [11, 12, 168] the residual ferrous and ferric oxide plays a role similar to magnesium oxide in pyroxene, that is it is one of the constituents of olivine. From the aspect of the determination of rock compositions a more realistic picture is obtained if the FeO and Fe_2O_3 in excess to the iron content of magnetite and ilmenite, respectively, is converted in the calculation to magnesium oxide, than when they are completely omitted from the composition.

From the oxidic analytical data of the rock the amount of ferric oxide bound by titanium dioxide and the quantity of ferrous oxide bond by the residual ferric oxide may be calculated. Examination of the thin sections of crystallized synthetic stone have confirmed that in the presence of ferric oxide, ferrous oxide forms primarily magnetite, since though the iron oxide content varies maximum between 1 and 2%, the presence of magnetite is clearly visible on the slices. The remaining FeO and Fe_2O_3 are converted into equimolar quantities of magnesium oxide which in the diagram will be included in the quantity of diopside representing the pyroxenes. If we wish to proceed more accurately, the magnetite and ilmenite contents must neither be left out of consideration. If the microscopic tests fail to reveal the presence of ilmenite, the TiO_2 content of the rock must also be converted into the stoichiometrically equivalent quantity of MgO. The data of the three-component system are thus complemented into a four-component system by the total quantity of magnetite and ilmenite and further transformed. By this method the behavior of the composition under investigation can be approached more satisfactorily, since this is an improvement of the procedure applied by ORMONT [11]. Diopside is the collective name for rhombic, monoclinic and triclinic pyroxenes which also include olivine.

The applicability of the basic principles of composition reduction is supported by the experimental proofs of CZWETKOW [2], LEONTEEV [168] and RISSE [7] who demonstrated the separation of ferric oxide and of an equivalent quantity of ferrous oxide in the form of magnetite from basalt melt.

Composition planning allows but for little iron oxide in crystallized synthetic stones, so that from this aspect no significant composition reduction is required. The smaller quantities of barium oxide are calculated as calcium oxide and the manganese oxides as magnesium oxide. Though the oxide composition of crystalline synthetic stone does not fit accurately the three-component system, from its silica and alumina content the area can be demarcated which characterizes the composition from the point of view of crystallization. The discrepancy is caused by the presence of higher quantities of $CaSiO_3$, wollastonite, further that part of the sodium oxide passes into the pyroxene.

From this method of representation and evaluation the different behavior of certain Hungarian basalt deposits from that of rocks usually applied in the preparation of molten silicates may be deduced. The composition corresponding to the apices of the albite–anorthit–ediopside three-component system is shown in Table 8, changes in the concentrations of the five oxides in Fig. 62. The parallel lines are the equi-concentration lines, the main equi-concentration lines intersect in point P. The dashed line is the eutectic line plagioclase-diopside.

Hence the quantity of sodium oxide is in one of the apices – albite – 11.82%, that of magnesium oxide – diopside – 18.62%, and in the other two apices 0%. It follows from the construction of the asymmetric diagram that the quantity of the respective oxide decreases evenly from the apex

TABLE 8

Oxide composition in the apices
of the albite–anorthite–diopside system

Oxide	Ab	An	Di
Na_2O	11.82	—	—
MgO	—	—	18.62
CaO	—	20.16	25.90
Al_2O_3	19.44	36.64	—
SiO_2	68.74	43.20	55.48

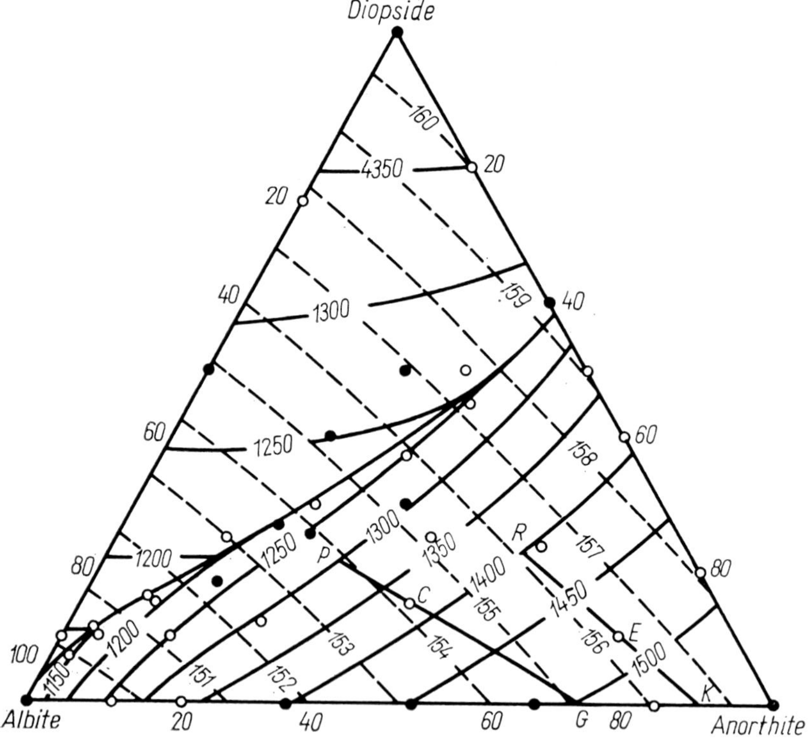

Fig. 61. The equilibrium conditions and liquidus isotherms of the albite–anorthite–diopside system

representing its maximum quantity to that where its quantity is zero (Fig. 63).

The quantities of calcium and aluminum oxides are different in two apices and zero in one, the quantity of silica is not zero in any of the apices, but varies between 43.2 and 68.7%. These data indicate that the diagram may be used between limits given by the values in the table. The principle may be applied to the investigation and comparison of other materials used in the silicate industries.

The basic triangular diagram may be subdivided into smaller triangles by drawing lines parallel to the sides of the triangles whereby a finer division is obtained. If each side is subdivided into 10 parts, divisions easily applicable to the characterization of silicate compositions are obtained (Fig. 64).

For the plotting of the main equi-concentration lines valid for apices *Ab*, *An* and *Di* the apex has to be joined to point P on the opposite side, in the case of apex D with B_x on the line $Ab—An$, where the concentration value of the given component is determined by the quotient

$$\frac{B_x - C_x}{A_x - C_x}$$

where A_x is the concentration of the investigated oxide on the albite apex; B_x and C_x are the concentrations on the anorthite and diopside apices, respectively.

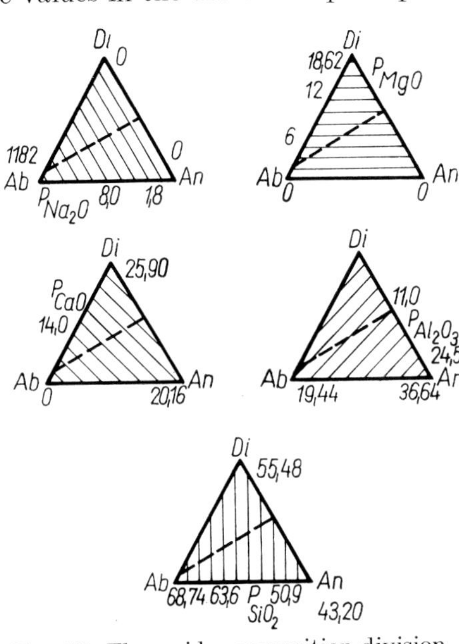

FIG. 62. The oxide composition division of the asymmetric state diagram derived from the albite–anorthite–diopside system

With the help of the equi-concentration lines the oxide concentration pertaining to any point may be determined; but for the demarcation of a reduced rock composition in the diagram the quantities of two oxides have to be known.

For the determination of the location of the rock the direction of the equi-concentration lines may serve as a basic indication. The location of the rock in the diagram gives the percentage of albite, anorthite and diopside, respectively, in its composition, that is from the weight percentage composition the mineralogical composition can be derived by simple means. If the quantities of magnetite and ilmenite, further the olivine content as part of the pyroxenes are estimated partly by accounting for the ferrous oxide quantity and partly from the microscopic investigation of polished slides, the complete mineralogical composition of the rock can be determined. From its location in the diagram the crystallization process of the molten

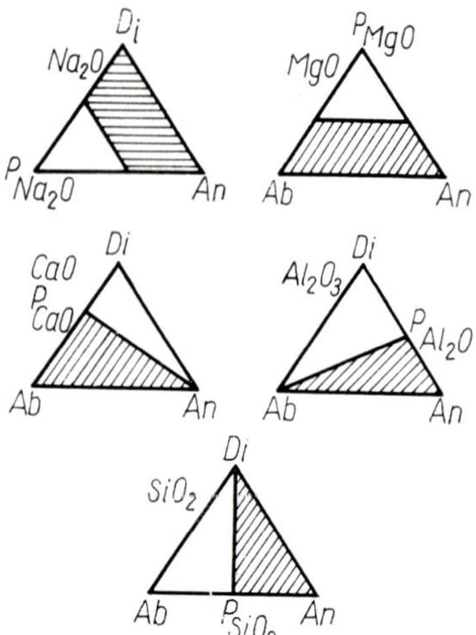

Fig. 63. Direction of the main equi-concentration lines of the oxides constituting the albite–anorthite–diopside system

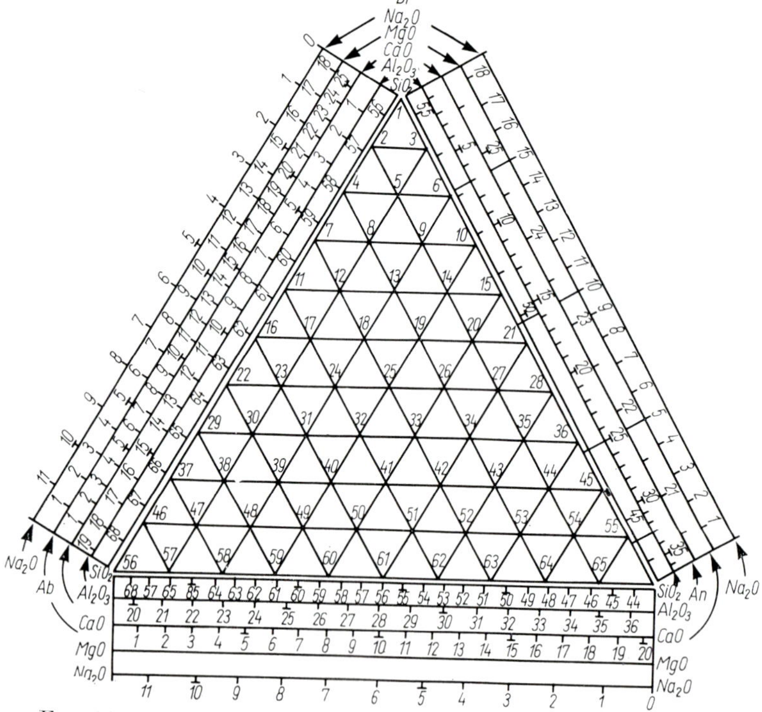

Fig. 64. The five oxide concentrations in the ternary system and the division of the base diagram into part triangles

rock can also be predicted. When finding the location of the rock the equi-concentrational lines usually do not intersect in one point of the diagram, but depending on the accuracy of the analysis will demarcate a smaller or larger area (Fig. 65). Beside experimental errors this may be caused by the presence of foreign substances [10, 11].

With the help of the diagram the following operations may be performed:

1. Determination of the oxide composition corresponding to any point on the diagram;
2. Determination of the mineralogical composition of rocks from the oxide composition marked on the diagram;
3. From their position on the diagram the qualification of rocks and molten silicates from the point of their melting and shaping behavior crystallization and processing parameters.

Investigation of rock compositions has shown that processable basalts and diabases are to be found in the areas Nos 1 and 2 of the diagram in Fig. 66. The composition of the crystallized synthetic stone may be assigned to area No. 3.

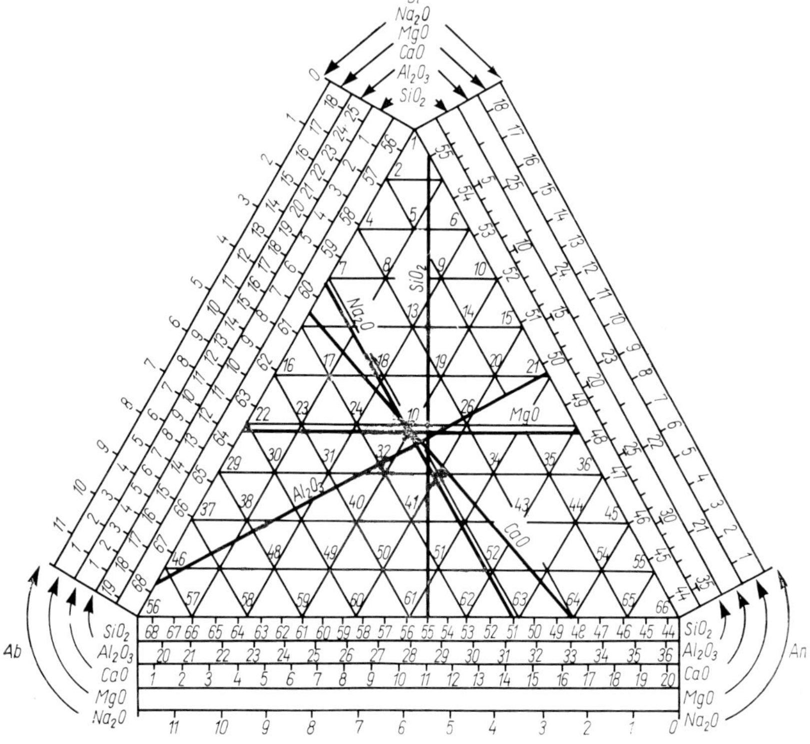

FIG. 65. Demarcation of the oxide composition from the direction of the equi-concentration lines in the phase diagram

It appears from the investigation of the magnetite-free reduced composition of rocks that the silica content varies between 50 and 60%. This value is between 50 and 55% for the reduced composition of easily processable rocks. Analysis of the pyroxene-plagioclase eutectic line and of the equi-concentration lines shows that since the equi-concentration lines of SiO_2, CaO and Na_2O are approximately parallel to the eutectic line marked by BOWEN, they fail to give a practical indication of rock behavior, while alumina concentration is considerably more characteristic. In the case of 0–15% alumina concentration the rock corresponds to the pyroxene field, at concentrations above 18% to the plagioclase field. Between these two extremes 15–18% characterizes the eutectic area. (The data refer to magnetite-free composition [1].)

The alumina content of practically processable rocks varies between 10 and 15% and preferably between 11 and 13%. The total quantity of alkaline earth oxides is about 28–31% of the reduced composition. It should be further noted that satisfactory processability is also characterized by about 10–13 parts by weight of magnetite to 100 parts by weight of rock, that is about 8–11% of magnetite in the rock.

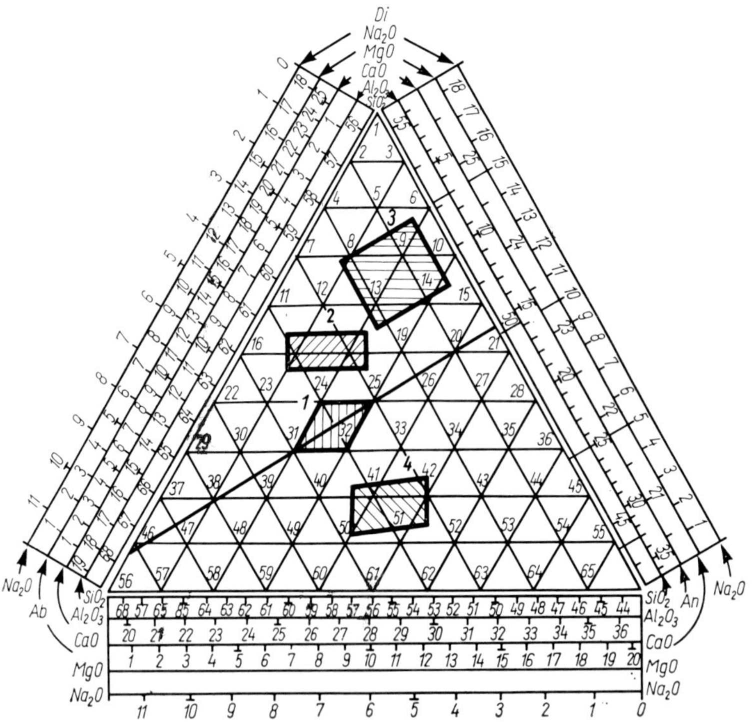

FIG. 66. The location of the composition of: molten rocks (1), (2), crystalline synthetic stone (3), Hungarian basalts (4) in the albite–anorthite–diopside system

Hungarian basalts contain either too much or too little magnetite and their alumina content usually exceeds the technically preferable 11–13%. The basalt deposits in County Nógrád-Gömör and in the Transdanubian region fall into area No. 4 in Fig. 66, thus into the plagioclase field.

With respect to the manufacturing technology of molten rocks we may also draw the conclusion that at least 8–13% of components with low liquidus temperature and a simultaneous viscosity reducing effect are required in the rock in which they will act as crystallization initiators. Magnetite and olivine contribute to this effect, but the quantity of olivine shall preferably not exceed 5% [1].

It was found that in general high olivine content results in a coarsely crystalline structure, that is in a material liable to cracking. The total quantity of feldspars and feldspar supplements shall not exceed 35–40%. Higher quantities of pyroxenes have a beneficial effect on rock processability, provided the pyroxene content is raised at the expense of feldspars.

It appears from the position of the crystalline synthetic stone in the albite–anorthite–diopside system that on the basis of its SiO_2 and Al_2O_3 content it is composed of 68–76% of pyroxene, 7–12% of albite and 10–20% of anorthite. The centre of area No. 3 in Fig. 66 corresponds to 72% of pyroxene, and 28% of plagioclase, while the basalt area No. 1 is characterized by 35% of pyroxene, 40% of albite and 25% of anorthite. The magnetite-free composition contains usually 30–40%, that is 47–54% of pyroxene to 68–76% of synthetic stone. The Hungarian basalts belonging into area No. 4 contain 15–25% of pyroxene [30, 31, 17]. The diabase deposit of Szarvaskő appears to be the most favorable (Table 4).

(C) Generalization of the production principles of vitroceramics

Generalization of the principle includes the demonstration of the nucleation effect of sulphide compounds, investigation of further nucleation effects in addition to those of manganese and iron sulphides, modification of the composition and development of a production method independent of furnace slag as base material.

For further studies of the nucleation effect the following base glass composition was used:

SiO_2	63.1%
Al_2O_3	13.8%
CaO	11.1%
MgO	7.2%
Na_2O	4.8%

This glass is characterized primarily by a higher SiO_2 and Al_2O_3 content, and consequently is more difficult to melt. With respect to its alkaline earth oxide content its CaO : MgO ratio is approximately the same as in dolomite.

The results obtained with the various materials used in the nucleation experiments are summed up in the following list:

1.	Base glass without additive	F, $+$, d
2.	Na_2S ($S = 3\%$)	F, B, d
3.	1.5% FeS	B, d
4.	1.5% FeS $+ 2.5\%$ MnS	K
5.	3% CoS	K
6.	3% Cr_2S_3	f, d
7.	3% Cu_2S	F, B, d
8.	3% MoS	K
9.	3% NiS	K
10.	3% Se	F, B, d
11.	4% MnSe	f, B
12.	2% F	O, B
13.	3.6% MnF_2	B, K
14.	3.6% $CaCl_2$	$+$, f, d
15.	4.2% $MnCl_2$	$+$, f, d
16.	3.6% MnS	K

where F is the surface crystallization with a $1-2\%$ relative increase in density. This is characteristic mainly of the base glass and shows at the same time a lack of nucleation effect; B is a nucleation process of fair quality with $3-5\%$ relative increase in density; K stands for a good nucleation effect with $5-7\%$ relative increase in density; F and B denote an intermediary character with concomitant surface crystallization and weak nucleation; d is the deformation in the course of heat treatment; f stands for a minimum surface creasing; $+$ for the absence of nucleation and O for the absence of crystallization.

The substance under investigation was added to the base glass and the experimental glass was melted at 1450 °C. The glass was worked into rods and disks and the liquidus temperature of the material, the maximum number of crystallization nuclei and crystallization rate were determined. The experimental materials were crystallized at the temperature of the maximum number of nuclei by giving the materials first a preliminary heat treatment of two hours in the temperature interval between 650 and 700 °C.

The experimental results are given in symbols representing the various states of crystallization. Thus beside iron and manganese cobalt, molybdenum and nickel were qualified as cations with good nucleating effect. It should be noted here that manganese as well as cobalt and nickel have two electrons in the outer N shell, while molybdenum has like silver one electron in the O shell.

Recently the nucleating role of the sulphide ion has been confirmed. In sample No. 2 crystallization began inside the material too, since, however, there is no manganese in the composition, and the quantity of iron is also low that is less than 0.5% due to the raw materials (sand, dolomite, feldspar) used, surface crystallization and deformation also occurred. The situation is similar in the case of selenium, but surface crystallization occurs even with MnSe indicating that Se has a weaker nucleating effect than S. The favorable effect of the sulphide ion appears also from its comparison with chloride, since when used in equal quantities crystallization is more favorable with sulphide than with chloride [170].

Generalization of the production principle is as follows: the main carriers of the nucleating effect are the heavy metal sulphides; the material can be based on feldspar–diopside, or wollastonite and enstatite. Analogue to the furnace slag composition finer structures may be obtained by adding low quantities of relatively many components.

Under the generalization of the composition we understand that independently of the raw material vitroceramics of widely varying compositions may be prepared on feldspar–diopside base by nucleation with heavy metal sulphide and having an alkaline earth oxide content.

The composition of the highly abrasion and chemical resistant materials varies between the following limits

silica	54–64%
alumina	5–17%
alkali metal oxide	2– 7%
alkaline earth oxide	10–30%
manganese oxide	1– 6%
iron oxide	0.5–4%

Sulphide content should be preferably at least 0.4–2.0% and in addition many other components may also be used, partly to improve the melting and molding conditions, partly to promote nucleation or to reduce crystallization rate. The necessary sulphide content is provided by the addition of calcium, magnesium, iron or manganese sulphide to the mixture. In practice it is preferable to reduce the metal sulphates such as sodium, calcium or manganese sulphate to sulphide. Reduction is best performed by the addition of coal dust, but other reducing agents may also be used. During melting the mildly reducing furnace atmosphere prevents the oxidation of the sulphide content.

Furnace slag behaves in fact extremely favorably as a base material for vitroceramics, since it contains besides the main components a number of constituents which have a highly beneficial effect on crystallization, thus in addition to the sulphides acting as nucleators it also contains TiO_2, P_2O_5, Cr_2O_3, and BaO. These last have also a very important role in controlling the crystallization tendency of the material. The few tenth of a per cent of fluorine content must neither be left out of the considerations. Table 9 provides information on the various base material compositions between the mentioned composition limits, while in Table 10 the compositions of the base materials are summed up.

To reduce the danger of deformation during crystallization nucleator combinations, that is several nucleating agents may be used concomitantly.

6

TABLE 9

Weight proportion of the base materials of vitroceramics
on feldspar–wollastonite–diopside base

	1	2	3	4	5	6	7	8	9	10	11	12
Aplite	50	50	50	—	—	50	30	30	25	16	30	30
Phonolite	—	—	—	31.5	31.5	—	15	25	25	16	25	25
Furnace slag	—	—	—	—	—	—	—	—	—	31	—	—
Na_2SO_4	5	6	6	2.5	3.3	—	3	—	2	3.5	1	3
$MgSO_4$	5	5	—	19.4	—	5	6	—	3	—	6	6
$CaSO_4$	—	—	2	—	2	—	—	6	6	—	—	—
Dolomite	38	48.4	51.4	—	48.3	28	10	10	—	—	38	10
Manganese ore	3	4	3	3	3	10	4	4	4	4	4	4
Talc	27.5	—	—	60.6	—	27.5	34	34	30	36	—	34
Sand	—	15.7	17.7	—	38.6	—	—	—	—	—	22	—
Coke dust	5	5	5	4	4	4	4	3	5	4	1	1

TABLE 10

The oxide composition in weight per cent of the base materials
of vitroceramics on feldspar–wollastonite–diopside base

	SiO_2	Al_2O_3	Fe_2O_3	CaO	BaO	MgO	MnO_2	Alk_2O
Aplite	77.5	12.6	1.2	—	—	—	—	7.0
Phonolite	57.6	20.6	3.6	1.9	—	0.3	—	13.1
Slag	36.7	7.0	0.9	41.6	1.2	5.5	6.1	0.4
Manganese ore	10.0	6.7	15.0	3.3	—	—	65.0	—
Dolomite	3.1	2.5	0.5	27.0	—	19.3	—	—
Sand	97.0	1.9	0.5	0.4	—	0.2	—	—
Talc	63.7	0.3	0.1	—	—	31.0	—	—

Thus for instance good results may be obtained by the use of molybdenum or cobalt compounds, or by nickel, molybdenum, tungsten or metallic silver or perhaps graphite.

LőcsEI has investigated the effect of various nucleating agents on model glasses. The model glasses belonged into the SiO_2–Al_2O_3–CaO–MgO–Na_2O system. The composition of the glasses is shown in Table 11.

The composition was modified according to the following system. The SiO_2, Al_2O_3, MgO and Na_2O contents were kept at the same level. The additives were added at the expense of the CaO content in the form of iron oxide, manganese oxide, phosphorus pentoxide and titanium dioxide, the first three in a quantity of 2% each, TiO_2, in a quantity of 4%. The sulphide was obtained by the reduction of sodium and calcium sulphate with coke. Fluorine was added in the form of AlF_3 by stoichiometric substitution.

From the glasses filaments, rods and disks of 5 cm diameter and 5–7 mm thickness were prepared.

The LITTLETON points of the glasses are shown in Table 12, together with the temperature where the glass filament begins to elongate under its own weight. Due to the progress of crystallization it was impossible to determine the LITTLETON point of some of the glasses.

TABLE 11

Composition in weight per cent of glasses (I–XXIV) used in nucleation studies

Components	I	II	III	IV	V	VI	VII	VIII
SiO_2	56	59.5	56	56	56	56	56	56
Al_2O_3	7	7.3	7	7	7	7	7	7
CaO								
MgO	32	28	28	28	26	26	26	28
Na_2O	5	5.2	5	5	5	5	5	5
Fe_2O_3	—	—	2	2	2	2	2	2
MnO	—	—	2	2	2	2	2	2
P_2O_5	—	—	—	—	2	2	—	—
Coke	—	—	—	—	—	—	6	6
F	—	—	—	—	—	—	—	—
TiO_2	—	—	—	—	—	—	—	—

Components	IX	X	XI	XII	XIII	XIV	XV	XVI
SiO_2	56	56	56	56	56	56	56	56
Al_2O_3	7	7	7	7	7	7	7	7
CaO								
MgO	26	26	26	26	28	26	28	26
Na_2O	5	5	5	5	5	5	5	5
Fe_2O_3	2	2	2	2	—	—	—	—
MnO	2	2	2	2	—	—	—	—
P_2O_5	2	—	—	2	—	2	—	2
Coke	6	—	6	6	—	—	—	—
F	—	3	3	3	—	—	3	3
TiO_2	—	—	—	—	4	4	4	4

Components	XVII	XVIII	XIX	XX	XXI	XXII	XXIII	XXIV
SiO_2	56	56	56	56	56	56	56	56
Al_2O_3	7	7	7	7	7	7	7	7
CaO								
MgO	24	22	24	22	22	24	22	24
Na_2O	5	5	5	5	5	5	5	5
Fe_2O_3	2	2	2	2	2	2	2	2
MnO	2	2	2	2	2	2	2	2
P_2O_5	—	2	—	2	2	—	2	—
Coke	—	—	6	6	6	6	—	—
F	—	—	—	—	3	3	3	3
TiO_2	4	4	4	4	4	4	4	4

In the base material of glasses VI, VII, VIII, IX, XI, XII, XIX, XX, XXI and XXII, 10% of Na_2O and CaO was present in the form of sulphates.

Crystallization conditions were studied by means of isothermic heat treatment. The lower temperature limit of crystallization and the liquidus temperature of the experimental glasses were determined. The liquidus temperature was nearly the same for all these glasses, namely 1210–1230 °C, the lower temperature limit of crystallization was 720-750 °C. The base glass began to crystallize at 790 °C, of those with nucleator content

TABLE 12

Softening points of the experimental glasses

	I	II	III	IV	V	VI
Littleton point elongation 1 mm/min	799	814	786	778	784	783
Beginning of filament elongation 0.1 mm/min.	766	773	754	754	747	743

	VII	VIII	IX	X	XI	XII
Littleton point elongation 1 mm/min.	780	779	775	—	—	749
Beginning of filament elongation 0.1 mm/min.	735	729	732	696	710	700

	XIII	XIV	XV	XVI	XVII	XVIII
Littleton point elongation 1 mm/min.	804	794	—	—	783	785
Beginning of filament elongation 0.1 mm/min.	747	730	683	710	724	715

	XIX	XX	XXI	XXII	XXIII	XXIV
Littleton point elongation 1 mm/min.	779	781	772	—	752	—
Beginning of filament elongation 0.1 mm/min.	733	726	721	716	711	716

the fluorine containing glass began to crystallize at the lowest temperature.

The following sequence of operations was observed for crystallizing the experimental materials and to determine the nature of crystallization and the formed texture:

Heat treatment

600 °C	1 hour
700 °C	1 hour
750 °C	1.5 hour
800 °C	30 minutes
820 °C	30 minutes
860 °C	3 hours

Without nucleating agent the glass belonging in the $SiO_2-Al_2O_3-CaO-MgO-Na_2O$ system shows a typical vigorous crystallization starting from its surface. Iron oxide and manganase oxide additives reduce the softening temperature without interfering with the character of crystallization, crystallization tendency increases as confirmed by the microscopic test of various crystallization processes. P_2O_5 considerably reduces the crystallization tendency of the base glass without nucleator, and in its presence the character of crystallization is unaltered, starting from the surface. P_2O_5 has no nucleating effect on glasses in the $SiO_2 - Al_2O_3 - CaO - MgO - Na_2O$ system, contrary to the findings with borosilicate glasses to which P_2O_5 has been added.

The intensive nucleating effect of sulphides (FeS, MnS) has again been confirmed. Phosphorus pentoxide acts by producing a finer structure. The addition of 3% of fluoride greatly raises crystallization rate, but produces a coarser crystal structure than nucleation with sulphide. The effect is similar with glasses containing sulphide; it is interesting that the LITTLETON point of this glass could not be determined, the glass filament showed a slight elongation only at the beginning of crystallization and none as crystallization progressed, which indicates a decrease of deformation tendency. In this case P_2O_5 also reduced the crystallization rate.

Titanium dioxide has a nucleating effect on the base glass. This is particularly noticeable if the material is first pretreated at a temperature corresponding to the beginning of filament elongation, when the dispersity of primary crystal separation will be higher and the texture finer. P_2O_5 will in this case too reduce the crystallization rate of the base glass.

Titanium dioxide and fluoride together have a strong nucleating effect. Fluoride raises the crystallization rate, but when applied concomitantly with TiO_2 it will produce a finer texture than alone which may be explained by enhanced nucleation, that is a rise in the number of crystal nuclei. In the presence of titanium dioxide and fluorine the addition of iron and manganese oxides cause practically no change in the composition, only a decolorization is observed. Without fluorine iron and manganese sulphides, just as titanium dioxide display a lower nucleating effect. This is explained by the masking of the Ti ion by the sulphide ion which is more readily polarized than the oxide ion and thus the Ti ion may preserve at least the bulk of its lower coordination even in a lower temperature range. It was further found that to ensure the nucleating effect of TiO_2 reductive condition have to prevail during melting. Fluorine will enhance crystallization rate even in the presence of TiO_2 and metal sulphides. Here the crystallization tendency reducing effect of P_2O_5 is also manifest even in the fact that the LITTLETON points of glasses containing fluorine, TiO_2 and sulphide cannot be measured, but if the glasses contain in addition P_2O_5 the measurement can be performed. The microscopic test of the crystallized samples point also to a decrease in crystallization rate.

Two theoretically highly interesting results arose from these experiments, namely the crystallization tendency reducing effect of P_2O_5, further the reduction in the nucleating effect of sulphide and TiO_5 when the two agents are applied concomitantly.

The behavior of P_2O_5 may be explained with the Al_2O_3 content. Al_2O_3 in quantities of 2–3% is known to reduce considerably the crystallization tendency of glasses as demonstrated on the classical example of the Thüringen glass tube. Below 3–3.5% the Al^{3+} ion with its coordination number of four occupies a network forming position. At higher concentrations

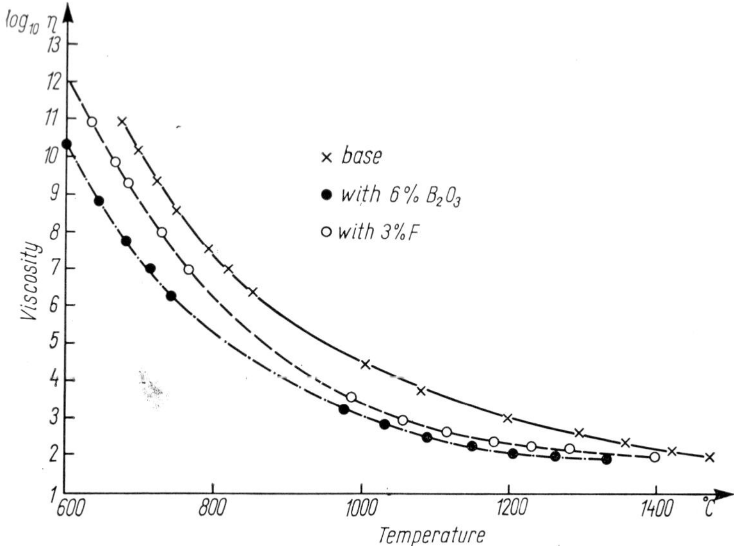

FIG. 67. Viscosity of the base glass of molten silicate type KM2

Al_2O_3 will have a coordination number of six and as such will act as a modifying agent when its higher quantities will contribute to the enhanced crystallization tendency of the glass. In the presence of P_2O_5 $AlPO_4$ double tetrahedrons are formed in the silicate structures, aluminum phosphate having a structure identical to that of quartz which explains the crystallization tendency reducing effect of P_2O_5 in high Al_2O_3 content glasses.

P^{5+} forming with Al^{3+} a double tetrahedron equivalent to two SiO_4^{4-} tetrahedrons is capable of transferring an equivalent quantity of Al^{3+} ions

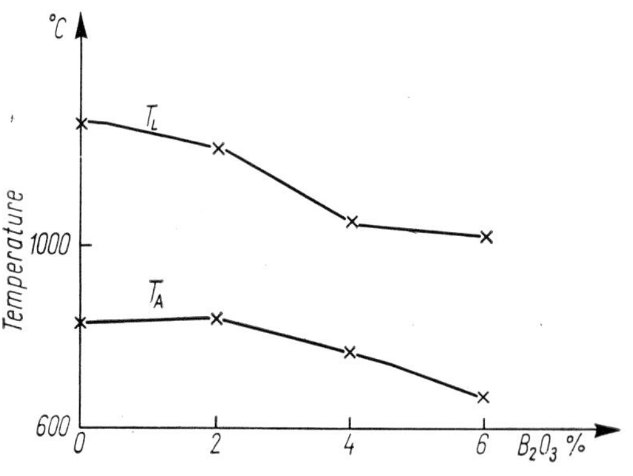

FIG. 68. Changes in the liquidus temperature of the vitroceramic type KM2 as a func tion of the B_2O_3 content

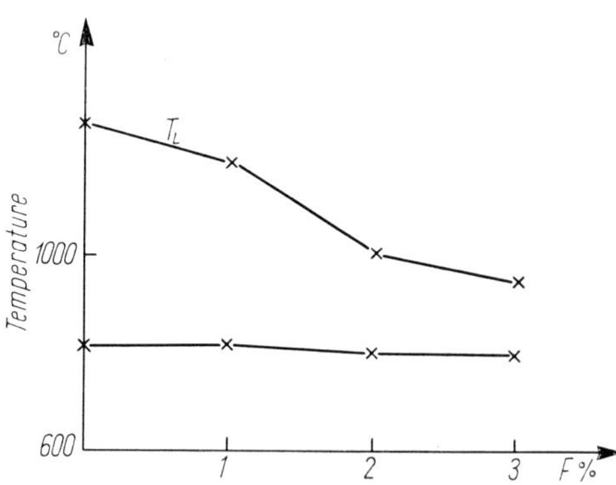

FIG. 69. Changes in the liquidus temperature of the vitroceramic type KM2 as a function of the F content

in the modifying position into the network forming position which results in an improved stability of the glass structure [148].

Several technological aspects have raised the possibility of reducing the liquidus temperature. In the glass industry this problem is solved mainly by the addition of boron trioxide and the analogy to hard glasses suggests here the same solution. In the experiments aimed at the reduction of the liquidus temperature the composition of the glass was as follows:

$$
\begin{array}{ll}
SiO_2 & 60\% \\
Al_2O_3 + ZnO & 16\% \\
MgO + CaO + BaO & 19\% \\
Na_2O & 5\%
\end{array}
$$

The viscosity vs. temperature curve of this glass is shown in Fig. 67. The glass has a liquidus temperature of 1260 °C with a lower limit of crystallization at 820 °C, crystallization rate reaches its maximum at 1150 °C, crystallization tendency at 1090 °C. In one series of experiments SiO_2 was substituted in the above composition by B_2O_3 in quantites of 2, 4 and 6% respectively. The addition of boron trioxide reduced the liquidus temperature and the lower limit of the crystallization area as shown in Fig. 68.

In another series of experiments fluorine was introduced in quantities of 1, 2 and 3% respectively without altering the cation content of the base glass. The effect of fluorine content on the liquidus temperature is shown in Fig. 69. Thus both boron trioxide and fluorine content reduces the liquidus temperature, while the lower limit of crystallization is reduced to a greater extent by B_2O_3 than as an effect of fluorine addition. Boron trioxide considerably reduces crystallization rate and the intensity of nucleus formation. Fluorine enhances the crystallization tendency, especially crystallization rate and nucleation tendency. The effect on the liquidus temperature is even more pronounced when boron trioxide and fluorine are used concomitantly.

From the model glass a vitroceramic material of the following composition was evolved:

$$
\begin{array}{ll}
SiO_2 & 60\% \\
Al_2O_3 & 15\% \\
CaO + MgO + BaO & 15\% \\
Na_2O & 5\% \\
MnO & 2\% \\
FeO & 1.1\% \\
Cr_2O_3 & 0.1\% \\
TiO_2 & 1.8\%
\end{array}
$$

The sulphates among the base materials are reduced to sulphide whose quantity is preferably between 1 and 2%. The material melts easily at 1400 °C and may be processed by molding under pressure or by casting. At a maximum temperature of 860 °C it is converted into a material of microcrystalline structure. Figure 70 is the light microscopic, Fig. 71 the

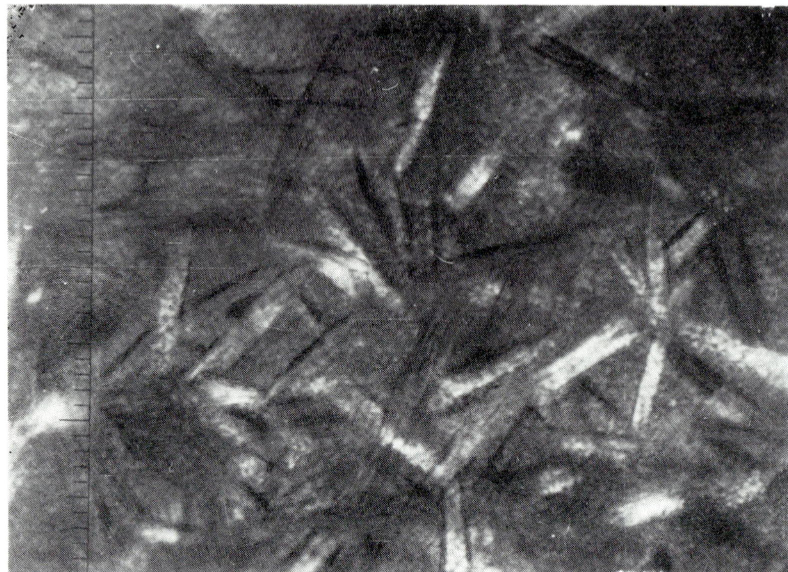

Fig. 70. Microscopic picture of a thin section of the vitroceramic type KM3 in transmitted polarized light. Magnification: 800×

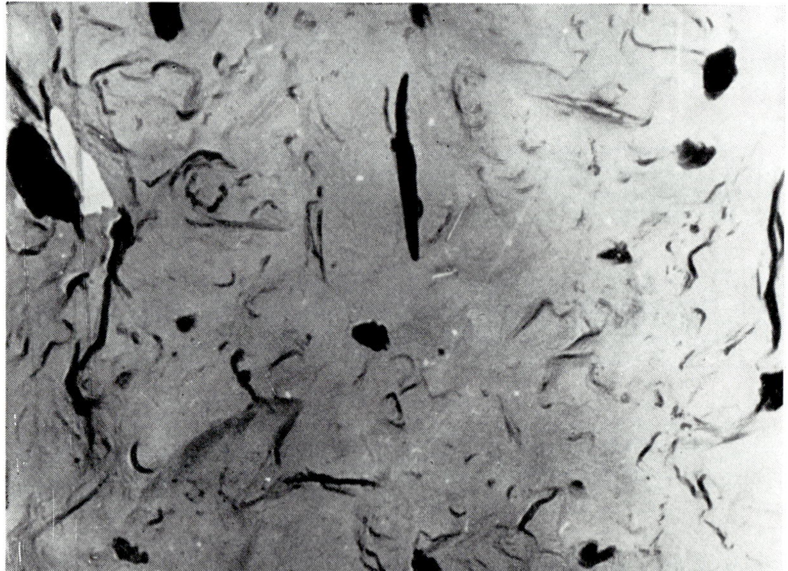

Fig. 71. Electron microscopic picture of the vitroceramic material type KM3. Magnification: about 24,000×

electron microscopic picture of this glass texture. The magnification of the light microscopic picture is $1500\times$, of the electron microscopic picture about $25{,}000\times$. The structure of this vitroceramic material consists of 10–15 μm long, 1.5–2 μm thick rods with albite crystals less than 1 μm large filling the voids between the rods. The electron microscopic picture of the fractured surface illustrates the crystallinity and fine texture of the material.

The liquidus temperature of the material is 1180 °C, the lower limit of crystallization is 780 °C. In the composition of this vitroceramic material SiO_2 was substituted by B_2O_3 and 3% of fluorine was added. To improve crystallization conditions micro-eutectic additive of 2% was applied to compensate the crystallization rate enhancing effect of fluorine. Fluorine caused the effect shown in Table 12. The softening point of the fluorine-free material could be determined, but after the addition of fluorine the LITTLETON point could no longer be measured, and a coarse macrocrystalline structure was observed in the pictures. The increase in crystallization rate could, however, be compensated by the micro-eutectic additive. The liquidus temperature of the modified vitroceramic material was 830 °C, its maximum crystallization temperature 700 °C. The structure after heat treatment is shown in Figs 72 and 73. Electron microscopic pictures of the fractured surface show a uniform and high quality crystallization.

Reduction of the liquidus temperature may be interesting from several aspects. In this way it was possible to implement the deposition of the material on metal which may considerably extend the application of vitroceramics.

Application of vitroceramics to metals depends on the following factors:

1. It should be possible to bake the base glass practically without crystallization in the temperature interval between 850 and 950 °C. In the experiments the material was baked on a metal coated with an enamel primer. The experiments indicated a LITTLETON point in the order of 600–650 °C as the condition of baking.

2. After baking the glass has to be subjected to a nucleating heat treatment in the temperature interval between 480 and 520 °C (approximately corresponding to a viscosity range between 10^{12} and 10^{11} poise) by a process appropriate to vitroceramics.

3. This heat treatment is followed by crystallization in the temperature interval between 700 and 800 °C.

The other condition for the utilization of vitroceramics as enamels is an adequate heat expansion coefficient. Experiments have shown that a heat expansion coefficient between 70 and 80×10^{-7} mm/mm °C will be necessary for the application of the vitroceramics to an enamel primer, for baking and for the adhesion of the coating to the metal.

Vitroceramics may be recommended as enamel coatings not only because of their highly favorable mechanical properties, but also because of their excellent corrosion resistance.

Adjustment of the thermal expansion coefficient was solved experimentally. Substitution of alumina by zinc oxide resulted in a higher heat expansion coefficient. In this way the sodium oxide content could be reduced,

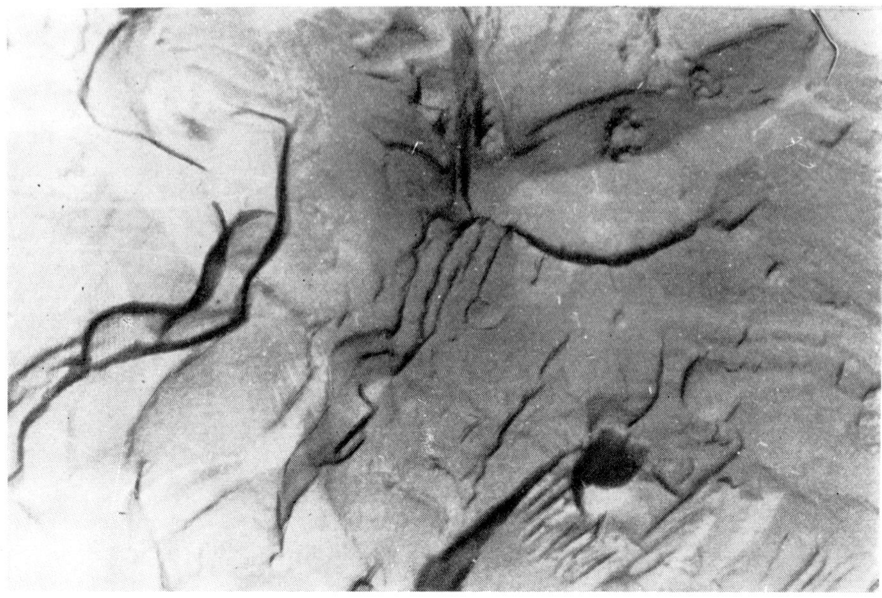

Fig. 72. Electron microscopic picture of a vitroceramic material with a liquidus temperature of 830 °C after 2-hour crystallization at 700 °C. The picture was prepared by means of carbon replica. Magnification: 28,000 ×

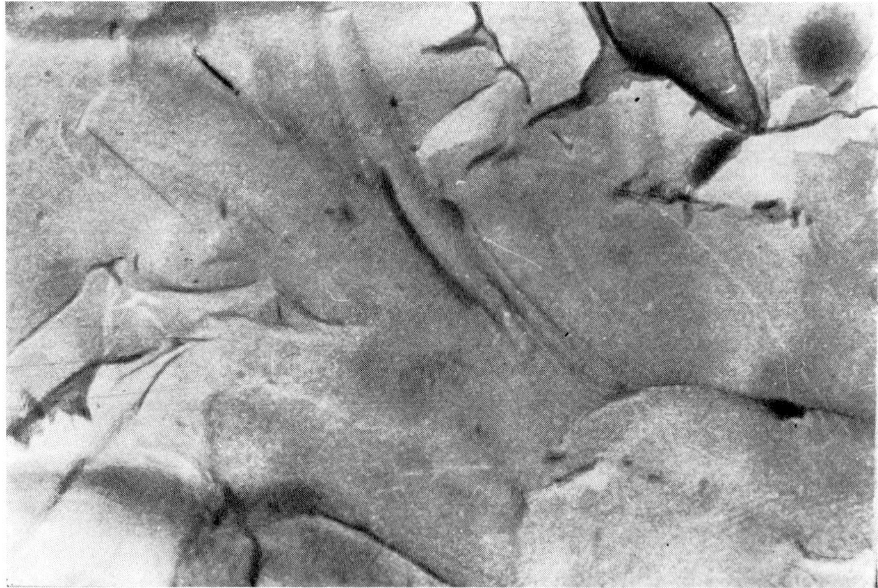

Fig. 73. Electron microscopic picture of a vitroceramic material with a liquidus temperature of 830 °C after 5-hour crystallization at 700 °C. The picture was prepared by means of carbon replica. Magnification: 16,000 ×

since under the given composition conditions sodium oxide is the most detrimental component from the aspect of acid resistance.

Experiments have shown that the material with the following composition meets best the conditions for enamel like applications:

SiO_2	52–66%
Al_2O_3	3–10%
ZnO	6–16%
TiO_2	2– 6%
BaO	2–10%
MgO	2– 7%
Na_2O	2– 9%
Li_2O	2–10%

Softening, that is baking temperature drops also under the effect of antimony oxide and lead oxide as parts of the composition. Metallic silver, cerium oxide, copper, cobalt and nickel oxide, each in quantities below 0.5% may be added as nucleating agents. To ensure crystallization and nucleation conditions a low quantity of fluorine and sulphide must also be contained in the composition. If a vitroceramic obtained by melting under mildly reducing conditions at 1400–1430 °C is fritted into water, ground with milling additives to small particle size and baked by the customary technology of the enamel industry, a coating may be obtained which resists 2 kg balls dropped from 80–90 cm height, a resistance several times higher than the 15–20 cm enamel strength value [171, 172].

Lower crystallization temperature has in addition a great influence on the economy of the process, when a reduction of the crystallization temperature by 150 °C has already a markedly favorable effect, especially from the aspect of the choice of the structural materials and lifetime of the heat treating kiln [170].

Chapter VII

Crystallization mechanism of vitroceramics on feldspar-diopside base

It follows from the effect of nucleators that the transition from the vitreous into the crystalline state occurs in two steps.

The first step is the separation of the nucleating agent, the second the heterogeneous catalysis and the crystallization process as determined by the crystallization tendency. Diffraction tests have shown that the primarily separating phase is the heavy metal sulphide and the microscopically identified chromium oxide.

On the basis of experimental evidence and theoretical considerations the following methods may be applied to ascertain the fact of crystallization:

1. light and electron microscopic tests
2. X-ray diffraction tests
3. determination of the thermal expansion coefficient
4. determination of the specific gravity
5. determination of the heat of dissolution
6. measuring of the flexural strength

In Fig. 74 the structure of the material KM2 in the vitreous state is illustrated by its X-ray diffraction pattern, in Figs 75, 76 and 77 by electron microscopic pictures. The pictures of the fractured surface of the brown glass show the microheterogeneous structure of the glass where the small dark spots are the coloring material. In Fig. 76 a more pronounced micro-heterogeneity of an initial phase separation nature may also be observed.

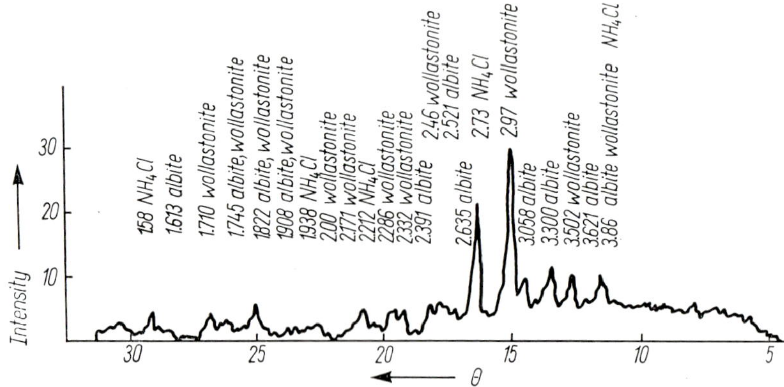

FIG. 74. X-ray diffraction pattern of a vitroceramic material type KM2 on feldspar–diopside base and nucleated with heavy metal sulphide

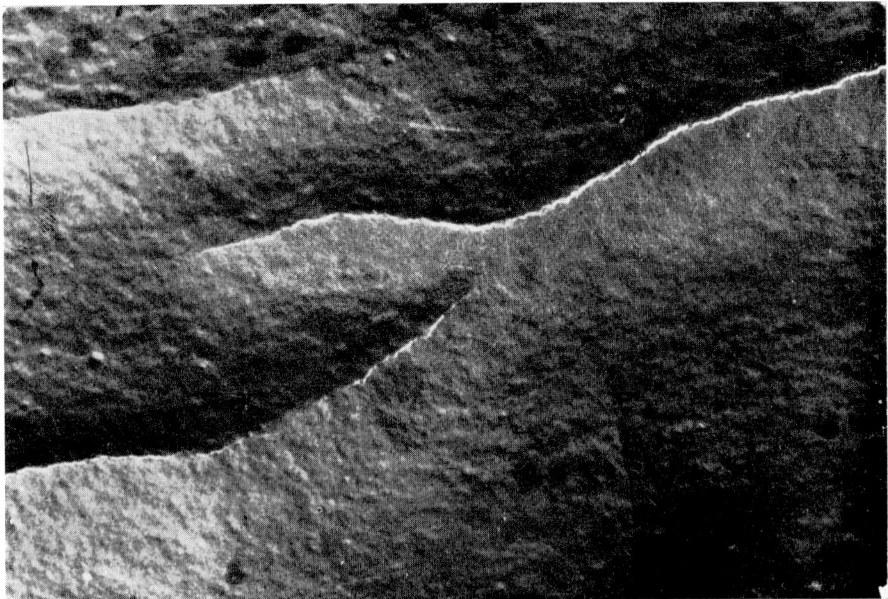

FIG. 75. Electron microscopic picture of the fractured surface of a vitroceramic material type KM2 cooled in the amorphous state. Magnification: 54,000 ×

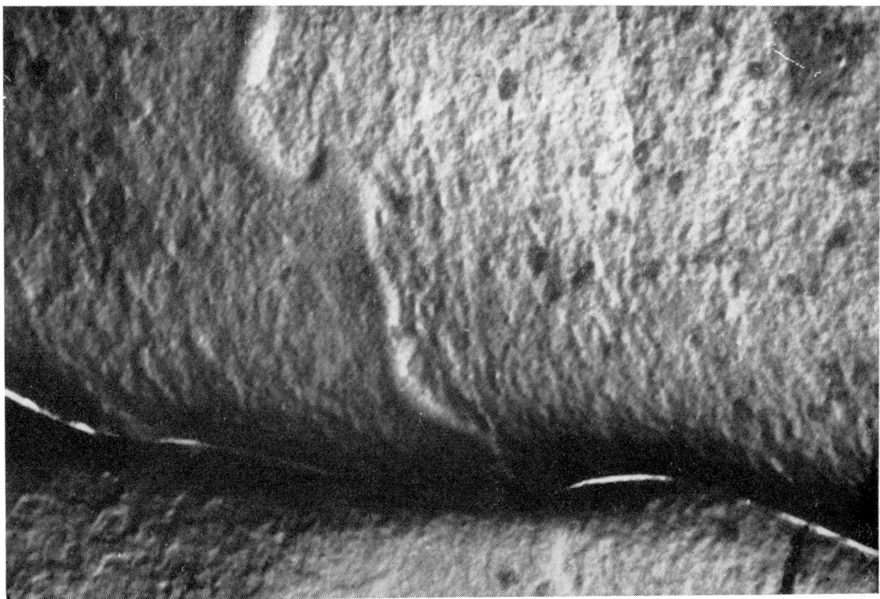

FIG. 76. Structure of a vitroceramic material type KM2 with signs of phase separation prior to crystallization. Magnification: 54,000 ×

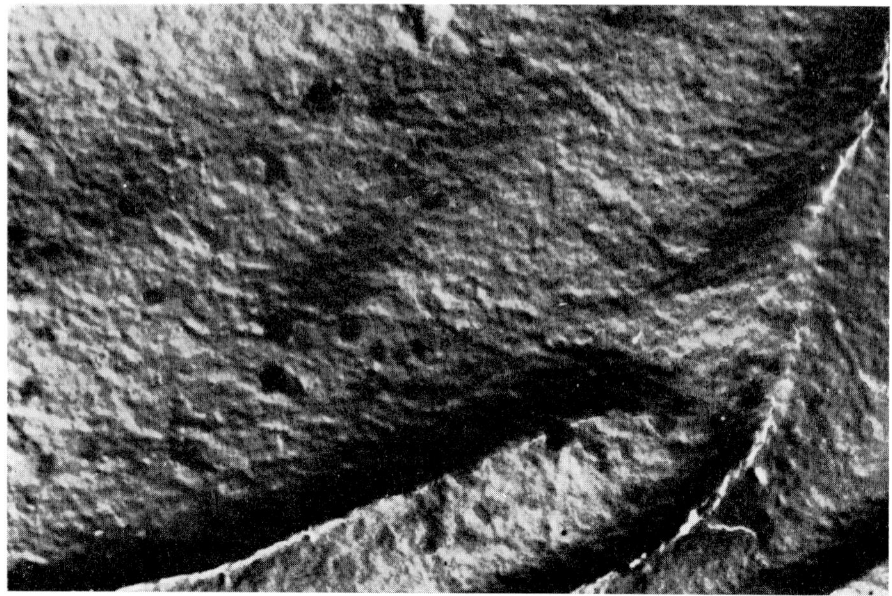

Fig. 77. Electron microscopic picture of the brown amorphous glass surface of a vitro-ceramic type KM2. The picture was prepared by means of carbon replica. Magnification: 54,000 ×

The liquidus temperature of the glass is 1265 °C, the lower temperature limit of crystallization without nucleation is 800 °C.

The material was heat treated in an electric furnace at 780 °C with a preliminary heat treatment for 3 hours. This was followed by heat treatment at 800, 850 and finally at 900 °C.

In the five-component system SiO_2–Al_2O_3–CaO–MgO–Na_2O nucleation, that is the separation of heavy metal sulphides begins in the vitreous phase under the action of heat treatment at 760 °C. At this temperature the separation of the nucleator is still very slow, but between 780 and 820 °C it proceeds rapidly. The kinetics of the process by the determination of the viscosity of the system becoming gradually heterogeneous, and by the increase in the quasi-viscosity value may be followed. In accordance with the kinetic nature of the process time will be the third parameter besides the temperature and viscosity. If viscosity is measured not statically, but dynamically by a method evolved specially for these investigations [20], nucleation will not be apparent because of the load on the glass (Fig. 78).

Glass No. I contained no nucleating agent. In glass No. II the nucleator replaces SiO_2 so that even the apparent viscosity will be lower than that of the base glass. In the figure the perpendicular in the $\eta - t$ (viscosity — temperature) coordination system represents the $\eta - \tau$ (viscosity — time) surface in which the effect of nucleation on the viscosity appears as a function of time.

In Fig. 79 the viscosity increase is plotted vs. the quantity of the nucleator. Viscosity was measured at the given temperature without load and after heat treatment.

$$\begin{aligned}
&\text{I.} && 1.5\% && \text{of nucleator}\\
&\text{II.} && 2\ \ \% && \text{of nucleator}\\
&\text{III.} && 2.5\% && \text{of nucleator}\\
&\text{IV.} && 3\ \ \% && \text{of nucleator}\\
&\text{V.} && 7\ \ \% && \text{of nucleator}
\end{aligned}$$

In all five cases the nucleators were iron and manganese sulphide, in glass No. V 2% of iron oxide and 5% of manganese oxide, in the other four two-thirds of iron oxide and one-third of manganese oxide were used. The sulphide ion originated from the reduction of sodium sulphate.

According to Fig. 79 increase in the quantity of nucleator approximately quadratically and time approximately exponentially raise the velocity with which the system turns into heterogeneous. Thus resistance to motion increases rapidly and the possibility of deformation decreases proportionally.

A certain difference appears in the nucleation intensity of crystallized synthetic stone and furnace slag free materials composed of pure oxides if the same quantity of nucleating agent is used. A certain fluctuation may be due to a small difference in reduction, but the statistical distribution of these deviations should be normal, while in the case under discussion the mixture containing furnace slag shows a definitely more pronounced nucleation.

This phenomenon may be explained by the fact that furnace slag contains in the order of a few hundredths of a per cent finely distributed elementary carbon, so that in the crystallization of "crystallized synthetic stone" every type of nucleating agent is represented. The nucleation intensity of a glass being furnace slag-free, but containing heavy metal sulphide can be considerably enhanced by the addition of vacuum sprayed Ni or W of not more than a few thousandths of a per cent. Figure 80 is the diffraction

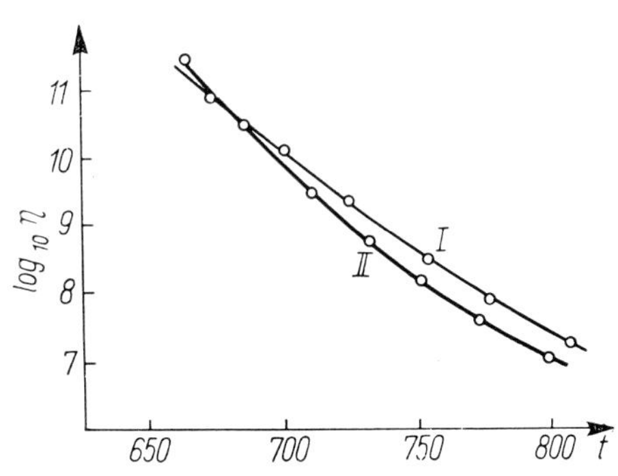

FIG. 78. Decrease in viscosity due to the addition of nucleators. I. Without additive, II. nucleator containing glass

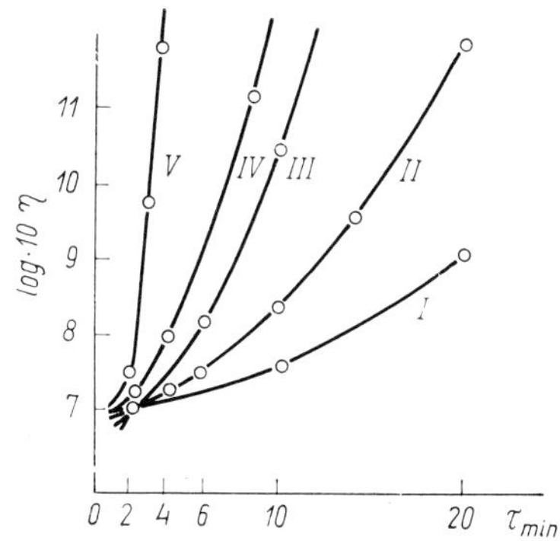

FIG. 79. The effect of increasing quantities of nucleating agents on the viscosity vs. the period of heat treatment

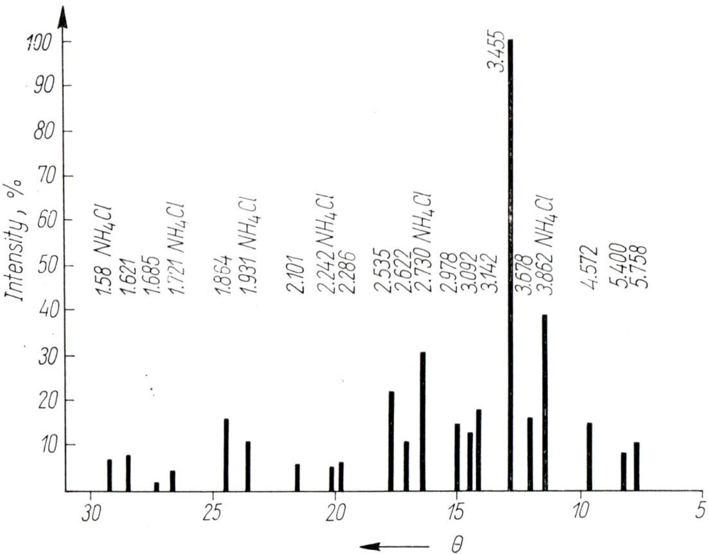

Fig. 80. X-ray diffraction pattern indicating the phases separating from a glass containing TiO_2, heavy metal sulphide and metallic silver nucleators and ZnO

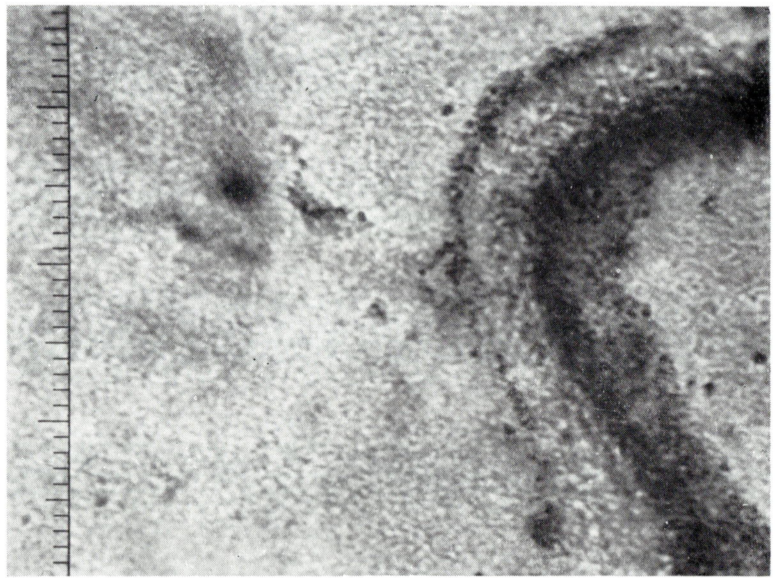

Fig. 81. Structure of a vitroceramic material prepared from furnace slag and rock in transmitted light. Magnification: 2000 ×

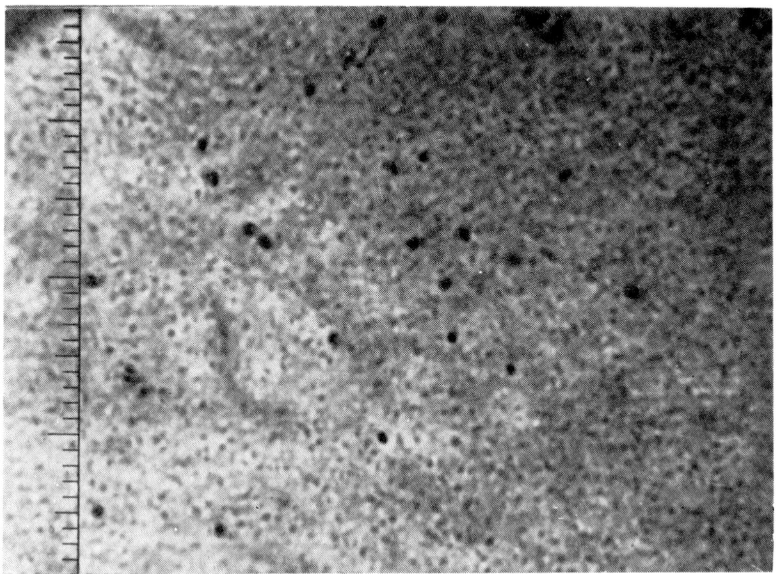

FIG. 82. Microscopic picture of a vitroceramic type KM2 (on phonolite–dolomite base) showing the uniform texture of the material. Magnification: 2000 ×

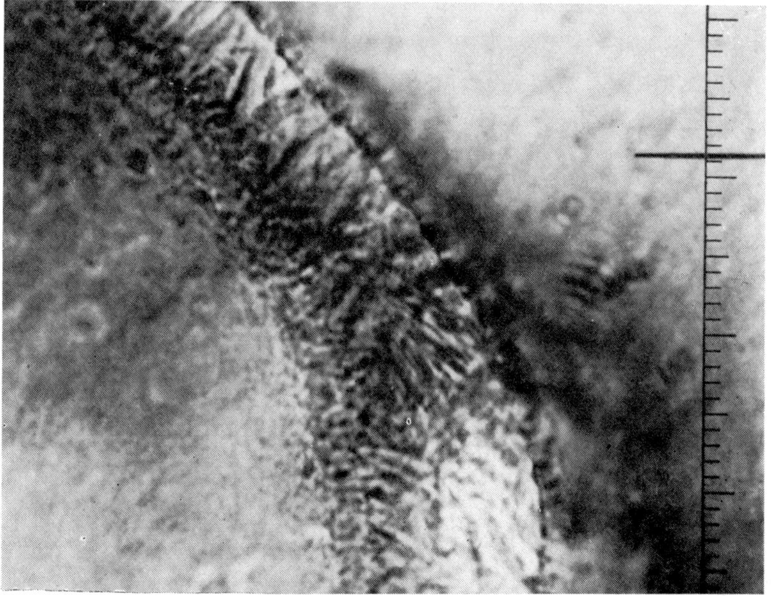

FIG. 83. Inhomogeneous part characteristic of the initial stage of crystallization, progressed state with low and high crystallization rate in polarized light. Magnification: 2500 ×

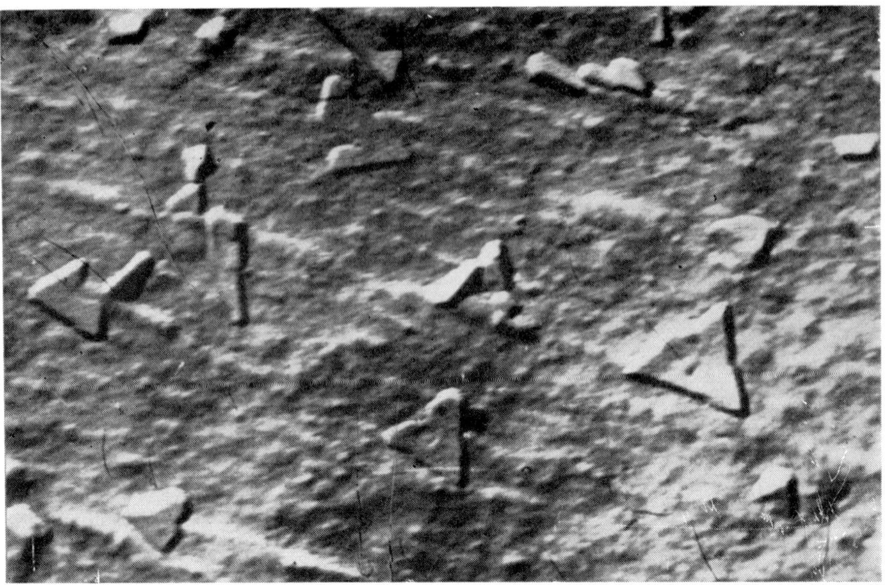

FIG. 84. The microphase separation phenomenon and nucleator separation under the effect of heat treatment at 780 °C. Magnification: 54,000 ×

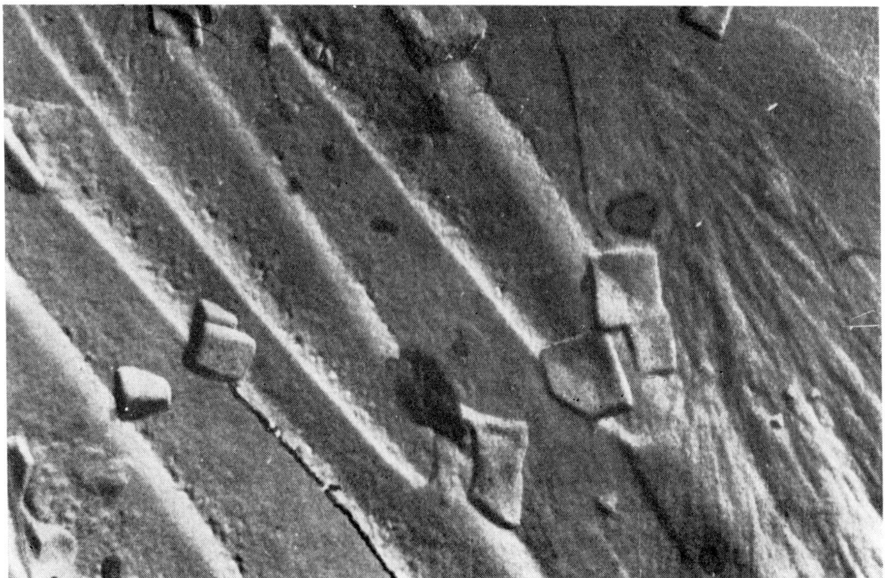

FIG. 85. Nucleator separating from the base glass of KM2 under the effect of heat treatment at 780 °C. Magnification: 54,000 ×, carbon replica

pattern of a vitroceramic on lithium aluminum silicate base containing ZnO (X-ray diffraction is the same as that of ignited petalite).

Pictures of the microsections of a sample after heat treatment (Figs 81 and 82) show that the texture consists of very small, 1–2 μm individual crystals.* Figure 83 is the light microscopic picture of the initial phase of crystallization.

The progress of crystallization is demonstrated on electron microscopic pictures in a 54,000$\times$ magnification. Figures 84 and 85 show the effect of heat treatment at 780 °C with the phase separation areas characteristic of vitroceramics, indicating the separation of the nucleating agent. This is the manifestation of the appearance of individual crystal nuclei which on the evidence of light microscopic and diffraction patterns consist mainly of iron and manganese sulphide.

Figures 86 and 87 show the structures after heat treatment at 850 °C for 4 and 8 minutes, respectively. The samples were first subjected to treatment at 780 °C for 1 hour. The beginning of crystallization is clearly visible, phase separation progressed and the structure became more heterogeneous.

The results of heat treatment at 850 °C for 30 minutes are shown in Figs 88 and 89, indicating a fairly varied structure with manifold structural elements. In 30 minutes crystallization has progressed quite significantly. Figs 90, 91 and 92 show after a preliminary heat treatment of 2 hours at 780 °C similar to that in the preceding experiment, the structural changes occurring at 850 °C. Figure 90 is the picture of a yet relatively slightly crystallized part, while Fig. 91 shows the structure of an already more crystalline surface element. The structures illustrated in Figs 92 and 93 are characteristic of vitroceramic materials. These pictures in themselves confirm that the mechanism of sulphide nucleation is of the same nature as nucleation of vitroceramics by means of metals and titanium dioxide.

Figures 94, 95, 96 and 97 show changes in crystal structure as a result of heat treatment first at 780 °C, followed by treatment at 850 °C for 120 minutes and at 950 °C for 180 minutes.

The crystalline phase of the product consists mainly of wollastonite and feldspar which is complemented in the proportion of the magnesium oxide content with diopside. The two forms of diopside are shown in Figs 98 and 99. The first is leaf-like shaped, the other is rod shaped. Figure 100 is the cross-section of these crystals perpendicular to their longitudinal axis and shows that the rods are empty inside. The form in Fig. 98 separates at low temperature, the crystal modifications presented in Figs 99 and 100, at higher temperatures. The well-developed crystals in Figs 98–100 separate from melts of lower viscosities and with higher alkali oxide content [106]. At dimensions of 0.5–1 μm it is of course impossible to identify the true form of the individual crystals. Heat expansion coefficient decreases with the shifting of the CaO : MgO ratio in the direction of MgO. To ensure satisfactory melting compositions with higher MgO content must also contain a certain amount of fluoride. Fluoride greatly accelerates crystalliza-

---

*The microscopic pictures were made by Zoltán Veress († 1965)

Fig. 90. A more poorly crystallized structural element of the material type KM2 heat treated for 120 minutes at 850 °C. Magnification: 54,000×, carbon replica

Fig. 91. Base structure with more markedly crystallized parts (lighter structural elements) of the vitroceramic material type KM2 after 120-minute heat treatment at 850 °C. Magnification: 54,000×, carbon replica of the fractured surface

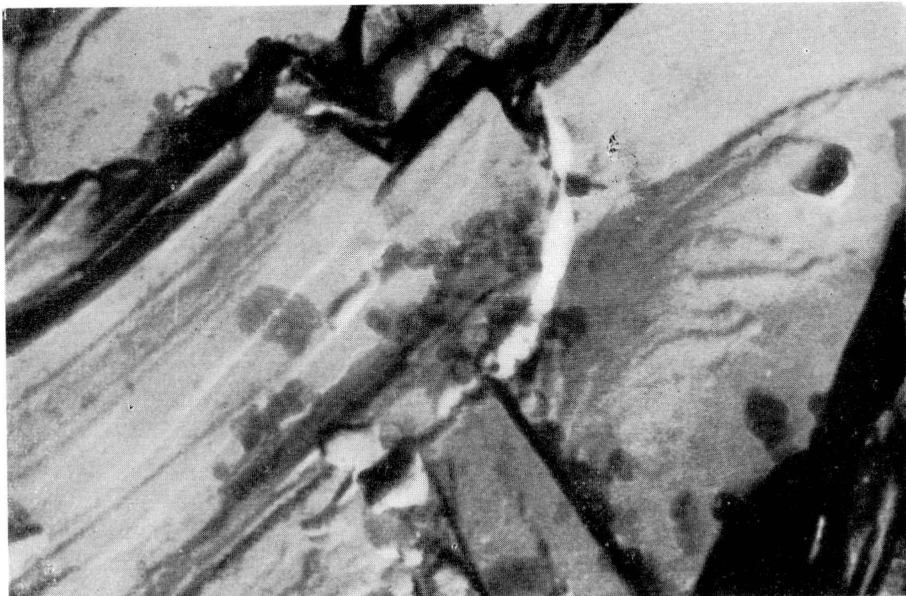

FIG. 92. The vitroceramic material type KM2 in a more progressed state of crystallization. Magnification: 54,000 ×

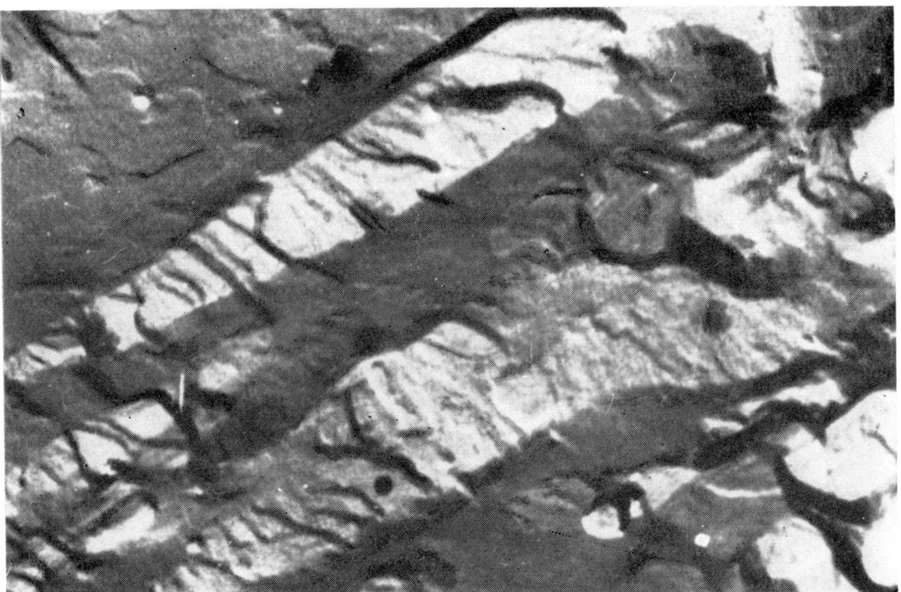

FIG. 93. Characteristic texture of the vitroceramic material type KM2 after crystallization at 850 °C. Magnification: 54,000 ×, carbon replica

Fig. 94. Characteristic texture of the vitroceramic material type KM2 after heat treatment first at 780 °C, followed by treatment at 850 and 950 °C. Magnification: 54,000 ×, carbon replica

Fig. 95. Characteristic texture of the vitroceramic material type KM2 first heat treated at 780 °C, followed by treatment at 850 and 950 °C. Magnification: 54,000 ×, carbon replica

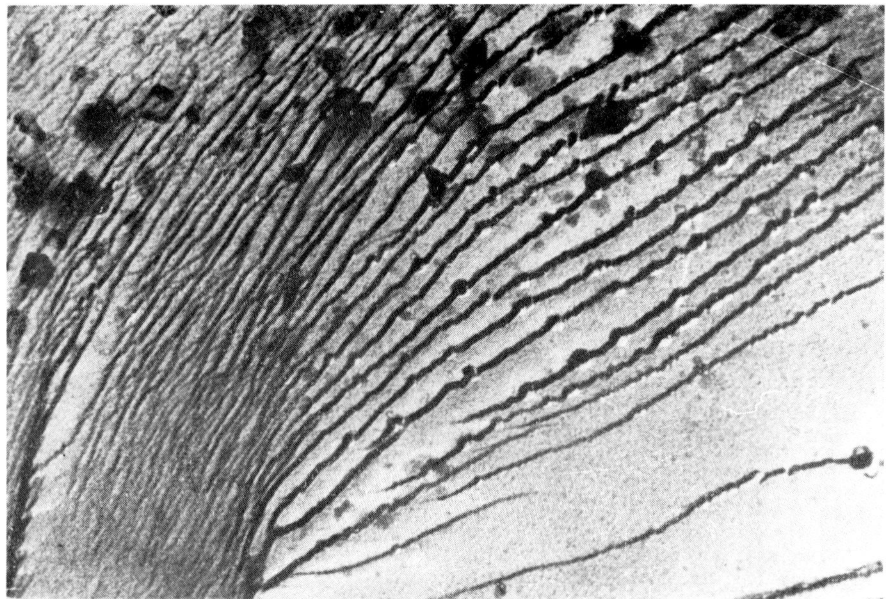

FIG. 96. An interesting part of the surface texture of the vitroceramic material type KM2 first heat treated at 780 °C, followed by heat treatment at 850 and 950 °C. Magnification: 54,000 ×, carbon replica

FIG. 97. Characteristic texture of a vitroceramic material type KM2 first heat treated at 780 °C, followed by treatment at 850 and 950 °C. Magnification: 54,000 ×, carbon replica

FIG. 98. Leaf-shaped formation, characteristically of diopside appearance formed in the presence of an amorphous phase; occurs in basalts. Magnification: 400 ×

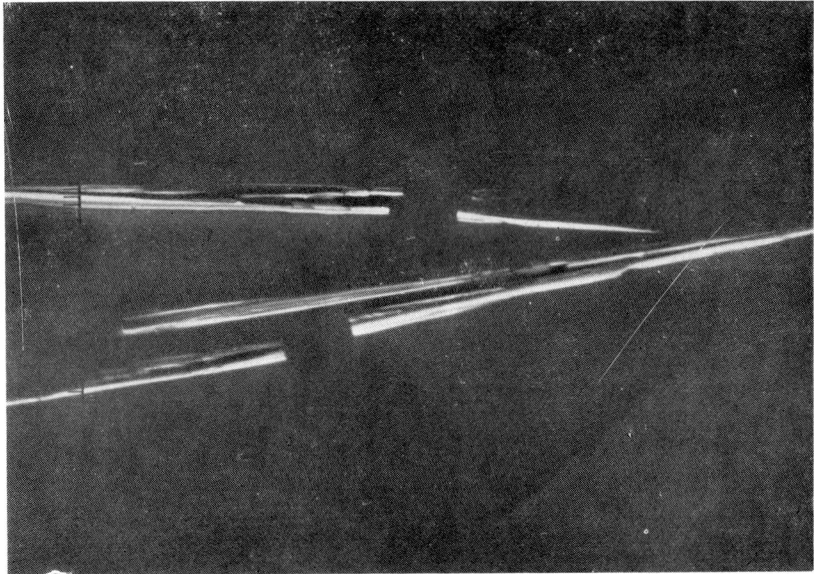

FIG. 99. Longitudinal section of the rod-shaped crystallization form of diopside. Magnification: 600 ×

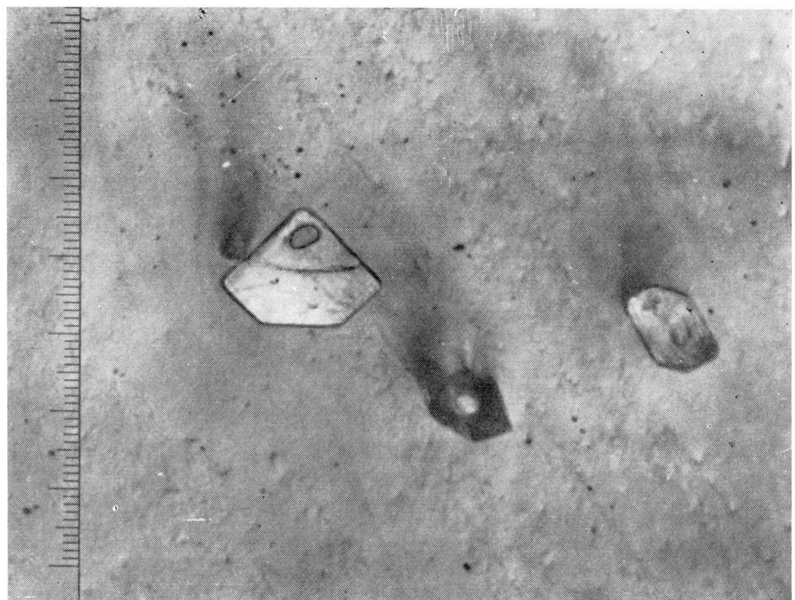

FIG. 100. Cross-section of the rod-shaped crystallization form of diopside.
Magnification: 1500 ×

tion, its effect may be balanced by the introduction of minor amounts of several new oxides.

The beginning of the crystallization process of vitroceramic base glass may be demonstrated on the DTA curve as shown by Lőcsei in Fig. 101. The curve was plotted at a heating rate of 40 °C per minute and gives a clear picture of the beginning of crystallization in vitroceramic materials of type KM2 [1].

Figure 102 shows in 1500× magnification the separation of the nucleating agent. DEBYE–SCHERRER tests have shown this separated phase to consist of iron and manganese sulphide. X-ray diffraction tests detected in addition the presence of silicates of complicated composition. The separation of the nucleator begins under the effect of a short heat treatment at temperatures between 760 and 780 °C.

Figures 103 and 104 provide further information on crystallization mechanism by showing the progress of crystallization in the material KM2 at magnifications of 500× and 1500×, respectively. The thin section in Fig. 104 demon-

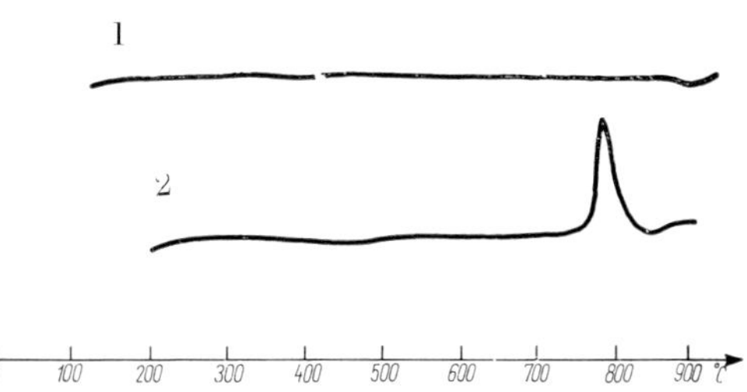

FIG. 101. DTA diagrams of amorphous and crystalline materials type KM2 on furnace slag base. 1. Crystalline synthetic stone, 2. slag glass

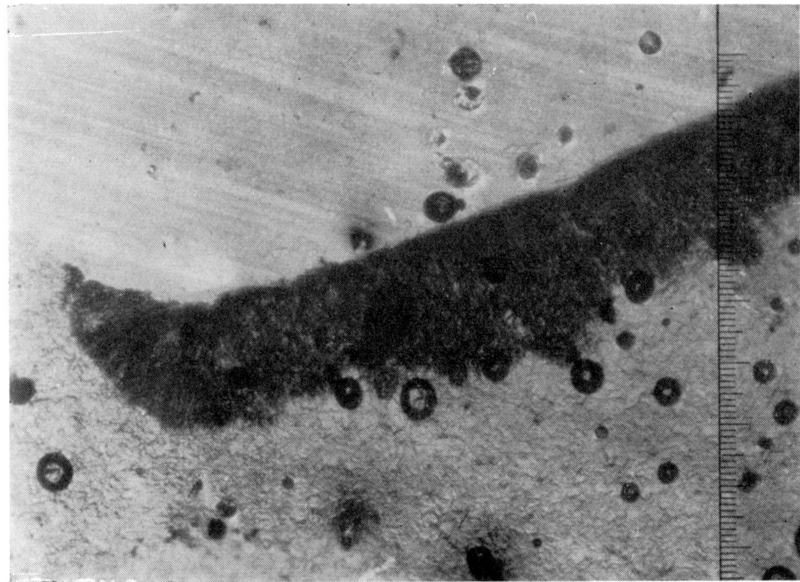

Fɪɢ. 105. Microscopic picture of crystallization beginning on the glass surface because of lack of nucleator. Magnification: 140 ×

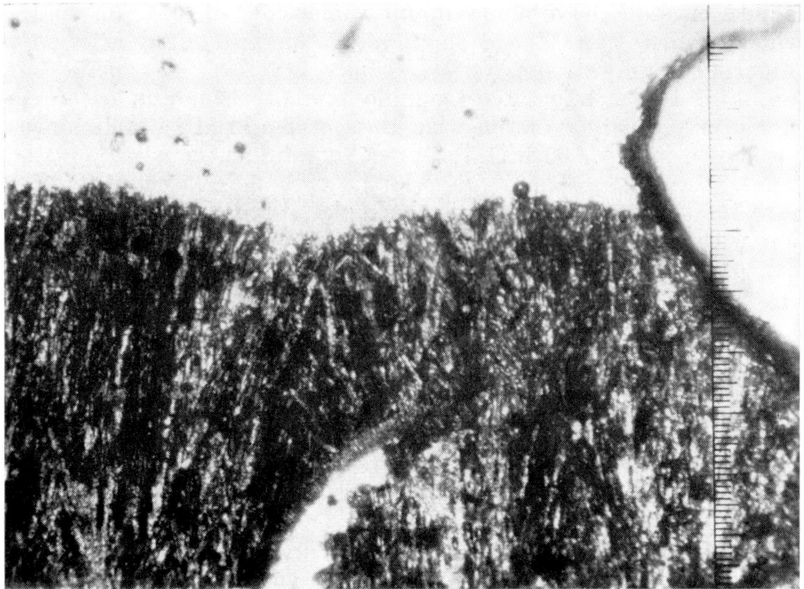

Fɪɢ. 106. Glass containing little fluorine. Because of insufficient nucleation crystallization begins at the glass surface. Magnification: 140 ×

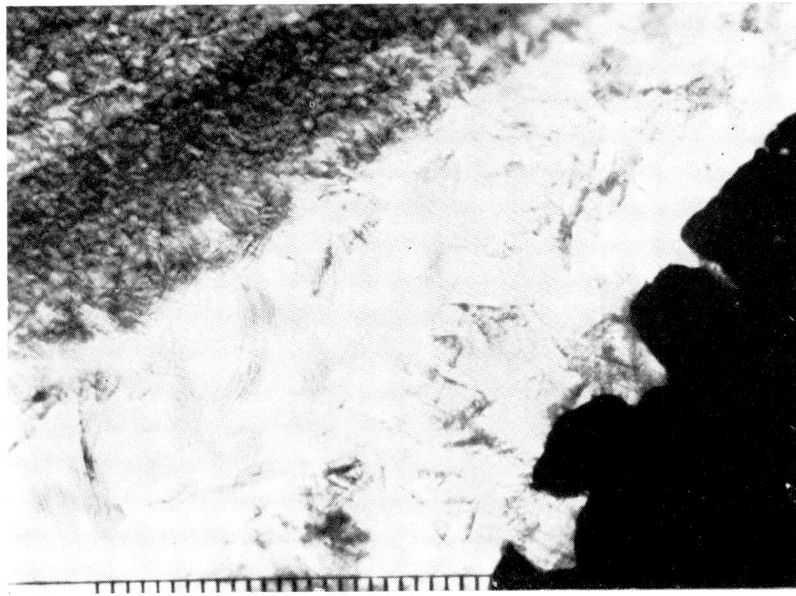

FIG. 107. Microscopic picture of the surface layer of the vitroceramic material type KM2. On the left: normal vitroceramic texture. Magnification: 400×

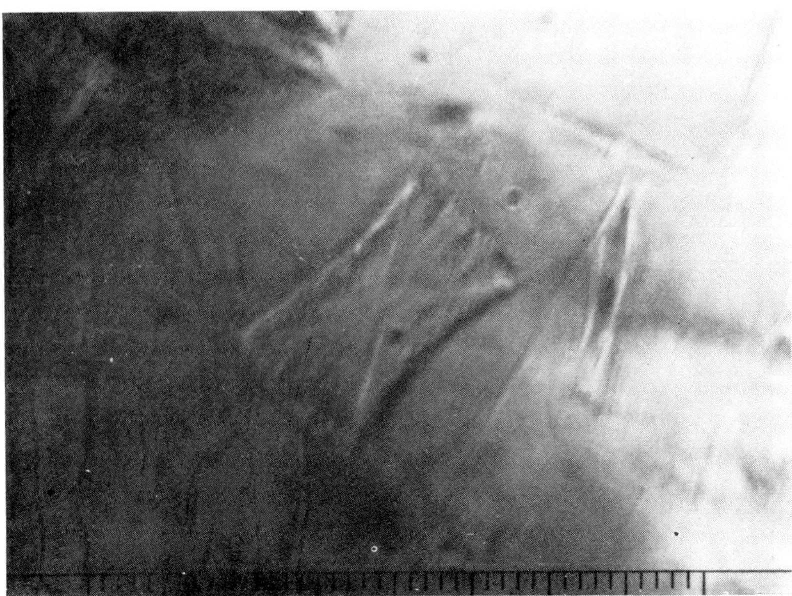

FIG. 108. Interesting textural element of the part with large crystals in Fig. 107. Magnification: 2000 ×

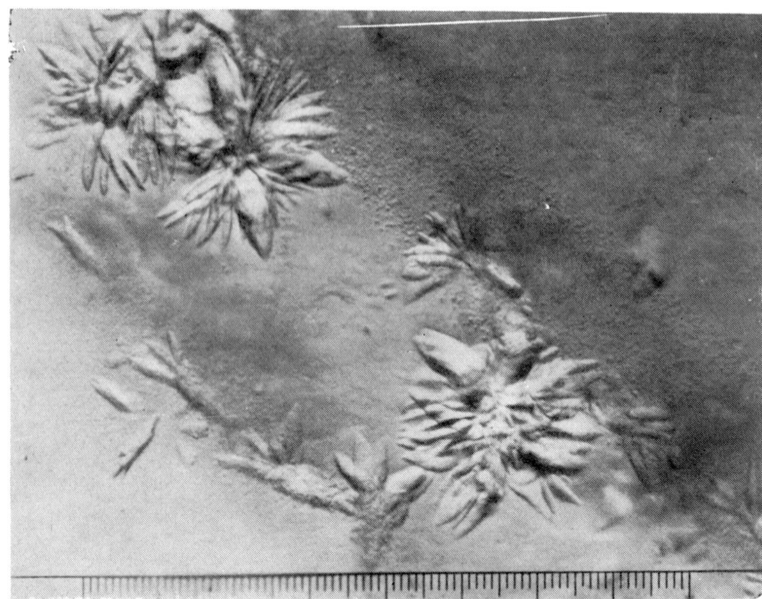

Fig. 109. Structural irregularities due to inhomogeneity in material type KM1. On the right: normal texture in transmitted light. Magnification: 1000 ×

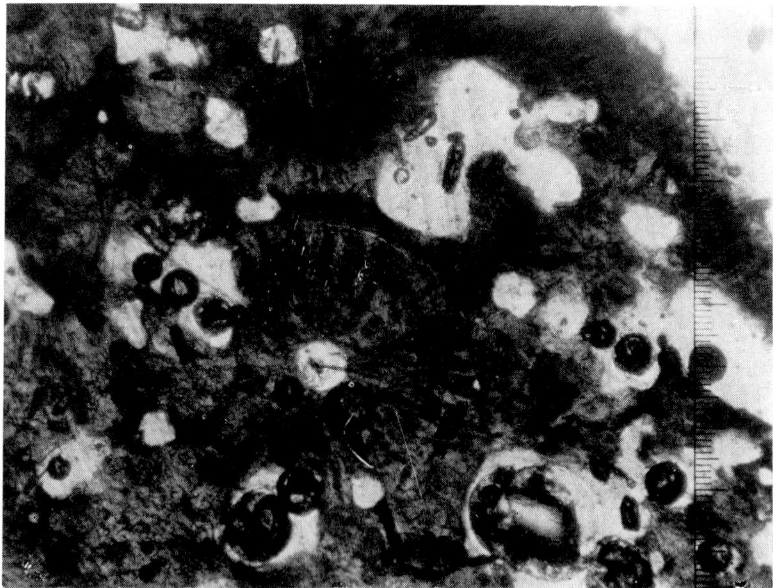

Fig. 110. Heterogeneous texture with large crystals in a vitroceramic material type KM2 brought about by the presence of fluorine. Magnification: 140 ×

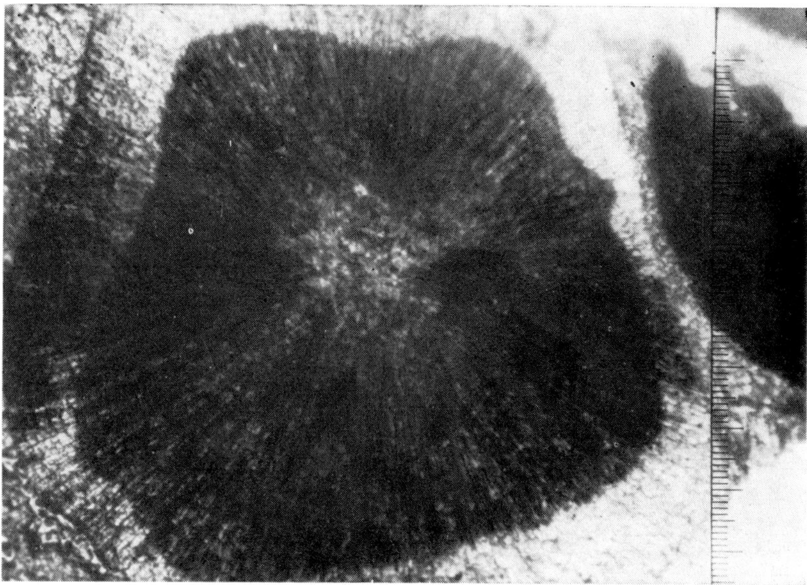

FIG. 111. Oriented texture formation in the vitroceramic material type KM2 under the effect of fluorine addition

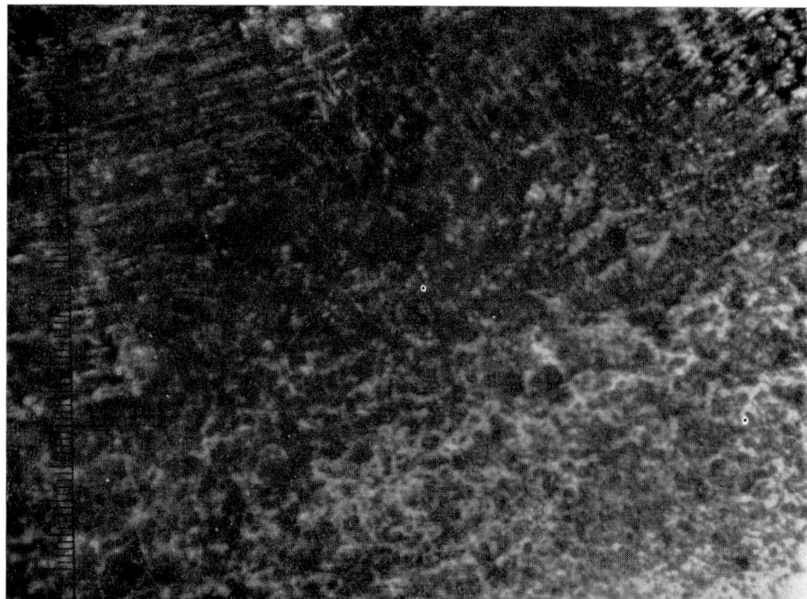

FIG. 112. Micro-eutectic additives (eight new components in a gross quantity of 2%) balance the crystallization rate enhancing effect of fluorine. Magnification: 240 ×

give a hay-stack like agglomeration in the heap. The individual filaments parallel to the side of the oblong give a straight extinction. The vibration direction of the higher velocity ray is parallel to the shorter side of the oblong.

Figure 109 shows large individual crystals beside the microcrystalline texture, these large crystals are due to inhomogeneous effects during the crystallization of the vitroceramic type KM2.

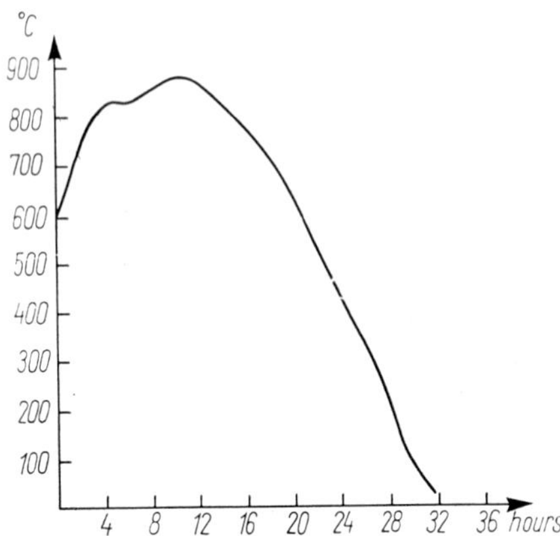

FIG. 113. Heat-treatment curve of the vitroceramic material type KM2

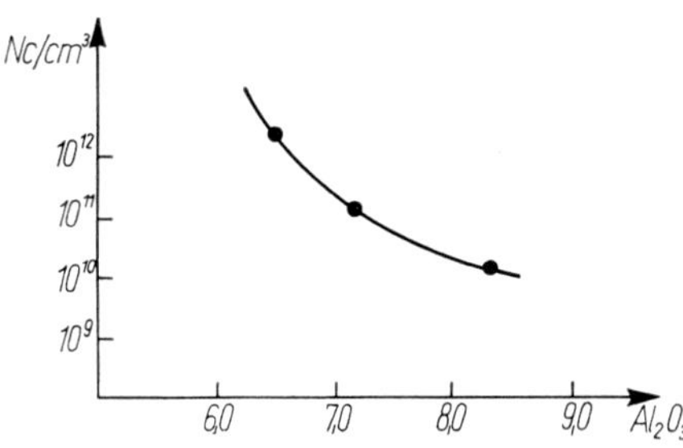

FIG. 114. The number of crystals (Nc) in the unit volume of the vitroceramic material type KM2 vs. the Al_2O_3 content

The microsection in Fig. 110 shows the significant increase in the crystallization rate of the vitroceramic type KM2 containing 2% of fluorine compared to that of a material of otherwise identical composition but without fluorine.

Figure 111 is the picture of a texture consisting of oriented crystals formed at high crystallization rates, a phenomenon which is also ascribed to the presence of fluorine in the composition.

The considerable decrease in the crystallization rate of vitroceramic type KM2 containing beside 2% of fluorine also micro-eutectic additive which appears to have eliminated the unfavorable effect of the fluorine content on the crystallization rate is illustrated in Fig. 112.

The heat treatment curve of the vitroceramic type KM2 under plant conditions is shown in Fig. 113.

Figure 114 illustrates the changes in the number of crystal nuclei in the vitroceramic type KM2 as a function of the alumina content indicating the considerable decrease in individual crystal sizes with decreasing alumina content.

In the case of 6–7% of alumina content the average crystal size in the vitroceramic type KM2 is 0.7–1 micron, from which it follows that for an average five-hour crystallization period the crystallization rate is 0.003 micron per minute. The same value for molten basalts is 7 micron, that is 1.4 micron per minute of crystallization rate, since in the molten basalt the average crystal size of 7 micron is formed within 5 minutes. Crystallization rate increases proportionally to the alumina content, at 8.5% of alumina content the crystallization rate reaches 0.02 micron per minute. Further increase of the alumina content will again result in a decrease in crystallization rate.

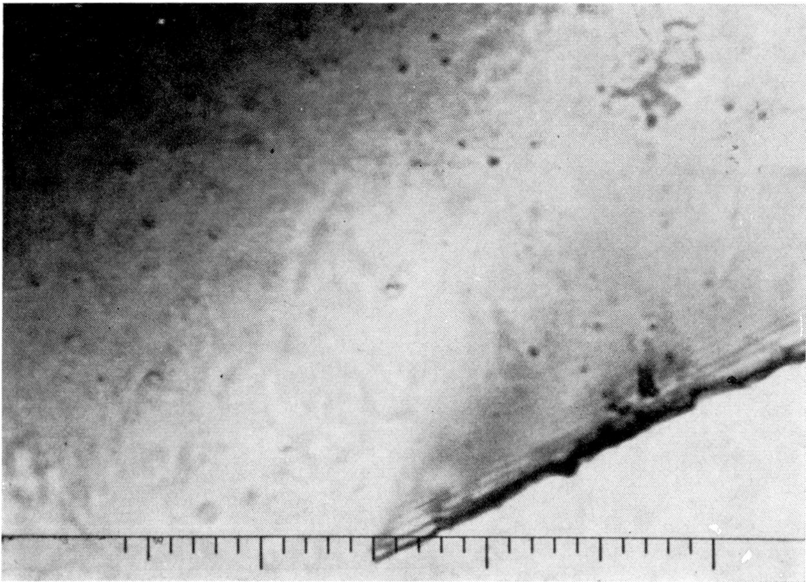

FIG. 115. Microscopic slide of glass KM2 after retarded cooling. Magnification: 3000 ×

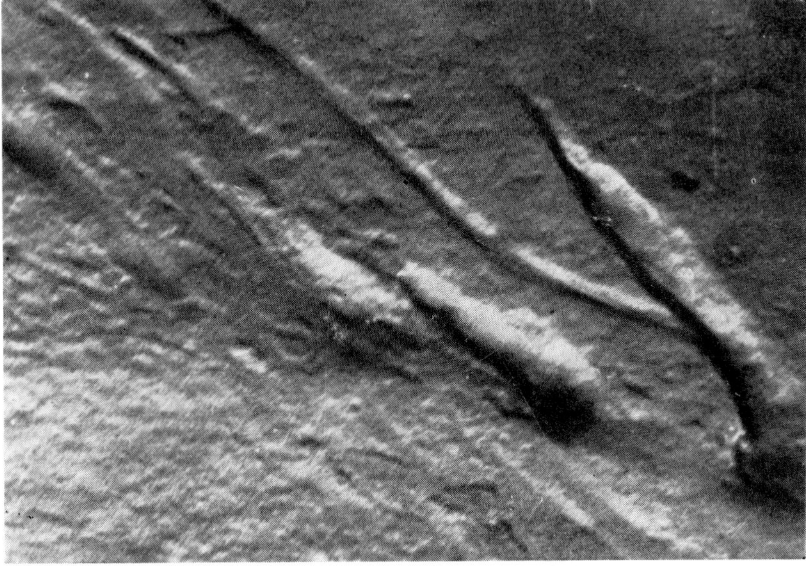

FIG. 116. Electron microscopic picture of the phase separation in material type KM2 as an effect of retarded cooling. Magnification: 54,000 ×

8*

Fig. 117. Phase separation and crystalline nucleator separation from amorphous vitroceramic material type KM2. Electron microscopic picture. Magnification: 54,000 ×

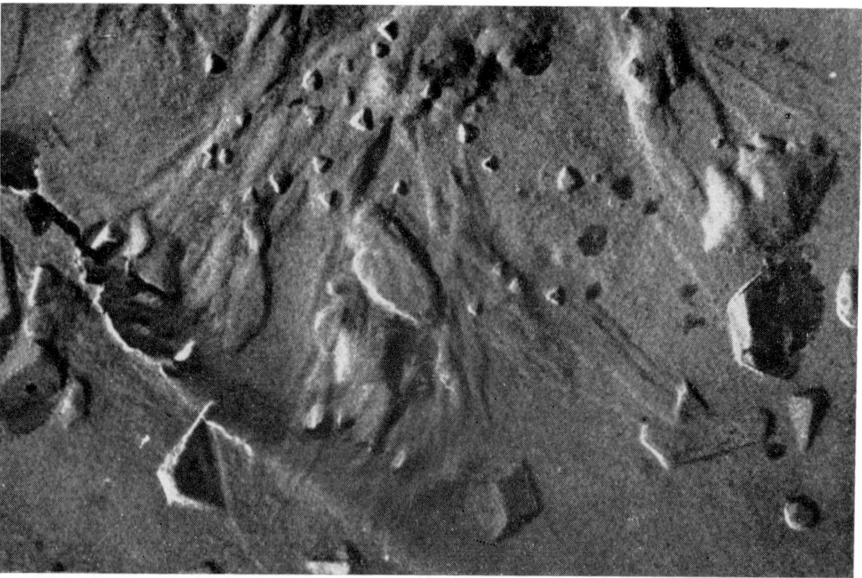

Fig. 118. A progressed stage of the phase separation process from amorphous vitroceramic type KM2. Separation of the heavy metal sulphide nucleator. Magnification: 54,000 ×

Fig. 119. Lamellar phase separation area in amorphous vitroceramic type KM2. A phenomenon characteristic of vitroceramics as an effect of retarded cooling

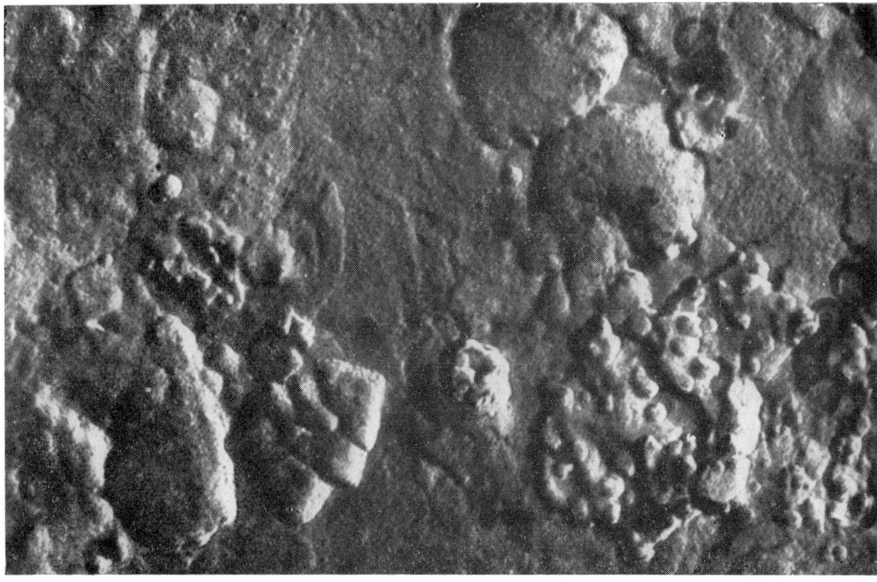

Fig. 120. Initial crystallization stage of KM2 glass as an effect of retarded cooling. Magnification: 54,000 ×

FIG. 121. Lamellar phase separation areas in amorphous vitroceramics type KM2 as an effect of retarded cooling. A phenomenon characteristic of vitroceramics. fication: 54,000 ×

FIG. 122. Initial crystallization mechanism of a vitroceramic type KM2. Progressed stage of phase separation. Magnification: 54,000 ×

Reduction of the cooling rate of the molten material will in the case of materials of the type KM2 already initiate the separation of the nucleator. When the molded melt is placed into a space of 700 °C temperature the microscopic picture shown in Fig. 115 will be obtained. In the bottom corner on the left larger Cr_2O_3 crystals are visible.

Figure 116 is an electron microscopic picture of the phase separation due to cooling. At $54,000\times$ magnification the structure prior to crystallization is clearly visible.

Figure 117 illustrates the phase separation phenomenon and the initial phase separation due to nucleation as a result of a brief heat treatment in the temperature interval between 750 and 800 °C.

Figure 118 illustrates a more progressed state of phase separation due to nucleator separation.

The electron microscopic pictures of the various stages of the crystallization process are shown in Figs 119–122 (magnification about $54,000\times$).

The progress of crystallization affects also the acid resistance of vitroceramics, so that the crystallization process may also be followed by acid resistance tests.

Figure 123 illustrates the acid resistance of amorphous KM2 after leaching with 10% hydrochloric acid.

Figure 124 shows the changes in the acid resistance of the material KM2 as function of the decreasing quantity of the vitreous phase.

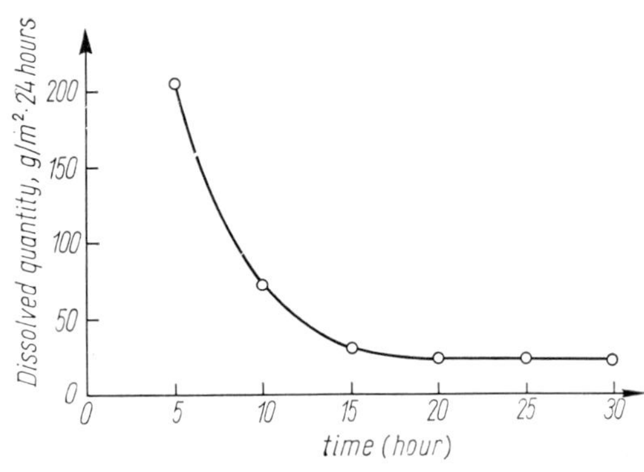

FIG. 123. Corrosion resistance of amorphous vitroceramic type KM2 towards 10% HCl at boiling temperature vs. the period of treatment

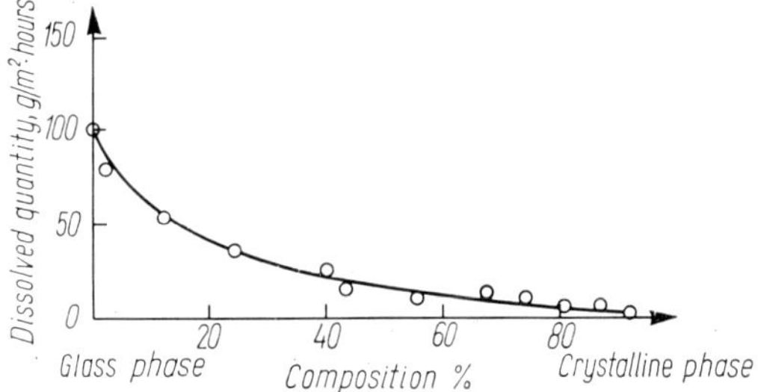

FIG. 124. Corrosion resistance of the vitroceramic type KM2 vs. the quantity of the crystalline phase

The tests lasted 8 hours in 10% hydrochloric acid. Improved acid resistance reflects the proportion of the crystalline phase to the amorphous phase in various stages of the crystallization process. The physical properties of the base glass will also undergo significant changes as crystallization progresses. In the amorphous state the heat expansion coefficient of the material KM2 is 88×10^{-7} mm/mm°C which by the end of crystallization drops to 76×10^{-7} mm/mm°C. In this way the progress of crystallization and the final state may be checked dilatometrically [106, 170].

TABLE 13

Changes in the corrosion resistance of material type KM2 vs.
the period of treatment

Reagent	Concentration %	Average dissolution $g/m^2 \times 24$ h	Dissolution in the first 24 hours g/m^2	Dissolution in the last 24 hours g/m^2	Dissolution rate $g/m^2 \times 24$ h		
					4 h	24 h	48 h
HCl	10	1.26	1.94	0.10	3.6	0.7	0.02
HCl	20	0.96	6.53	0.26	21.8	1.2	0.1
H_2SO_4	10	3.55	7.47	1.58	16.8	1.9	0.5
H_2SO_4	20	2.40	4.50	1.34	8.4	2.9	0.5
HNO_3	10	6.62	12.34	4.58	19.7	6.5	2.2
HNO_3	20	8.62	16.27	5.69	39.3	9.6	3.4
NaOH	10	7.99	13.97	3.22	36.7	4.1	0.2

Chapter VIII

Properties and possible applications of vitroceramics

The properties of molten silicates are determined partly by their chemical and mineralogical compositions and partly by their structure. The mechanical properties of vitroceramics belonging into the Na_2O–CaO–MgO–Al_2O_3–SiO_2 system and prepared by means of heavy metal, sulphide nucleation are primarily determined by the size of the crystals. As illustrated by Fig. 125 compression strength will be the more favorable the greater the dispersity of the crystals constituting the texture of the material. To an average individual crystal size of 0.7 μm a compression strength of 6000–8000 kp/cm² is assigned when the cube has a dimension of $40 \times 40 \times 40$ mm, and 10,000–12,000 kp/cm² with a cube dimension of $10 \times 10 \times 10$ mm. The flexural and tensile strengths measured on a quadratic prism of 160 mm length and 1600 mm² cross-section are 500–700 kp/cm², on an 80 mm long prism with 100 mm² cross-section 750–1100 kp/cm².

By improving the composition the flexural strength may be raised up to 1100–1400 kp/cm².

Improved strength values may be obtained also by reducing the size of the crystal forming the texture and by the adjustment of the quantity of the vitreous phase to an optimum value. Strength may be particularly improved by balancing the alumina content with the quantity of the other components. This improvement may be attained especially by the adjustment of the ratio of alkaline earth metal and alkali oxides to alumina. Flexural strength values up to 30–40 kp/mm² were obtained on circular diameter (of 4–5 mm diameter), 65 mm long samples which were supported at a distance of 50 mm. The quantitative determination of the optimum ratio of the crystalline to the vitreous phase is still lacking. The results of the softening point determinations were not sufficiently reproducible, so that conclusions with respect to the optimum crystallization time and period were drawn by an indirect method, just from changes in flexural strength.

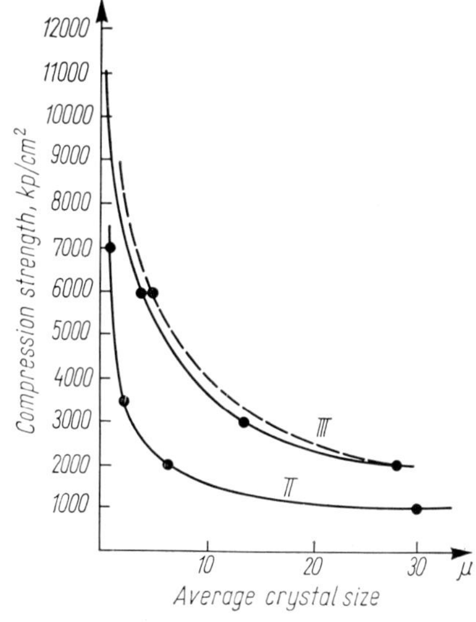

FIG. 125. Compression strength of molten silicates vs. average crystal size KM2 = II

The heat expansion coefficients of the three types of vitroceramics were as follows:

KM1	$110-120 \times 10^{-7}$ mm/mm°C
KM2	$65- 75 \times 10^{-7}$ mm/mm°C
KM3	$25- 45 \times 10^{-7}$ mm/mm°C

The hardness of the materials is $1000-1100$ VICKERS degree or $66-69$ ROCKWELL degree. Abrasion resistance is a complex property and changes depending on the type of wear and the quality of the abrasive.

Polishing tests gave the following results for the relative abrasion resistance of crystalline synthetic stones (the abrasion resistance of manganese steel against a high speed diamond disk was taken as 1; the greater the numerical values, the lower the abrasion resistance of the material)

	Factor
Pyrover laboratory glass	4.5
Steel crust	2.3
Badacsony basalt	2.9
Czechoslovakian molten basalt	1.9
KM1	1.4
KM2	1.3
KM3	1.1

The type of wear may, however, considerably influence the results. In the case of sliding abrasive wear the relative resistance of KM2 is 0.9, of KM3 0.75 compared to the resistance of manganese steel. In cement chutes the lifetime of KM2 was six times longer than of steel when subjected to the same wear.

Comparison between the acid resistances of molten rocks is extremely complicated, since there are no uniform test methods or ways to report the results. Evaluation of surface resistance in $g/m^2 \times 24$ hours seems to be the most expedient and is generally accepted in international practice. Thus resistance is described by the quantity in grams of the material dissolved from a 24 m^2 surface by different reagents in 1 hour. The test results should be converted to this dimension. Corrosion resistance to hot and cold reagents should be treated separately. There are three classes of corrosion resistance:

I.	from	0	to	2.4	grams per m^2 in 24 hours
II.	from	2.4	to	24	grams per m^2 in 24 hours
III.	from	24	to	240	grams per m^2 in 24 hours

The resistance of molten basalt in cold mineral acids is given in approximately the same way by French, German, Soviet, English and Czechoslovakian authors [2, 3, 7, 17, 19]. GINZBERG found that 27.4 g dissolved in concentrated HCl, 24.0 g in concentrated HNO_3 and 12 g in concentrated H_2SO_4 from a square meter surface in 24 hours. RISSE gives his results in weight per cent; he found that in concentrated HCl 0.013%, in dilute HCl 0.165% in concentrated HNO_3 0.001%, in dilute HNO_3 0.074%, in concentrated H_2SO_4

0.001%, in dilute H_2SO_4 0.503% was dissolved. According to PORTEVIN molten basalt is absolutely resistant to sulphuric acid and to cold alkalis.

Resistance to hot mineral acids decreases proportionally to the increase in temperature, though PORTEVIN gives no numerical results. According to GINZBERG 1.79% of the material is dissolved by boiling in hydrochloric acid for one hour, but he fails to mention the size of the surface. The following data are quoted for the sake of comparison: when a specimen of crystalline synthetic stone of 18 cm² surface and 9 g of weight is boned in 20% hydrochloric acid for 4 hours the loss of weight is 0.06%.

The results of the Czechoslovakian authors are the most reliable. They found that the dissolution of molten basalt in cold acid is 20–30 g/m²×24 hours, in boiling 20% hydrochloric acid 240–400 g/m²×24 hours.

Crystalline synthetic stone is absolutely resistant to cold mineral acids, of course with the exception of hydrogen fluoride. Its resistance to phosphoric acid is also limited. At room temperature 0.002–0.003 g of crystallized synthetic stone is dissolved from a 1 m² surface in 24 hours.

At boiling temperature the average acid resistance of crystallized synthetic stone is as follows:

in 20% HCl	12 g/m²×24 h
in 10% HCl	16 g/m²×24 h
in 20% H_2SO_4	16 g/m²×24 h
in 10% H_2SO_4	13 g/m²×24 h
in 20% HNO_3	30 g/m²×24 h
in 10% HNO_3	22 g/m²×24 h
in formic acid	24 g/m²×24 h
in acetic acid	12 g/m²×24 h

Series of acid resistance tests on the same samples pointed to a gradual passivation of crystalline synthetic stone, as testified by its decrease in dissolution to 0.5–1.5 g/m²×24 h after boiling for 20–30 hours. Solubility in alkaline salt solutions, e.g. in 10% sodium carbonate, also decreases rapidly and may be considered practically zero. Its resistance to alkalis is lower than to acids and the degree of passivation is not the same.

Chemical resistance is among others a function of the texture, the above data refer to a texture of average fineness. From the aspect of acid resistance it is an important fact that crystalline synthetic stones and molten rocks take up no water, their true porosity is less than 0.1%.

The corrosion resistance of the material KM2 is illustrated in Figs 126–129. The change in resistance as a function of time points to gradual passivation here too, towards both mineral acids and sodium hydroxide.

The figures show the quantity dissolved from unit surface in 1 hour and the total quantity of material dissolved per unit surface. The samples had a surface of 50–80 cm². The material Minelbit-1 possessed the highest resistance to hydrochloric and sulphuric acid, but its resistance towards nitric acid and sodium hydroxide is somewhat less favorable. Nevertheless it was found that there are but few corrosion resistant materials with an equally satisfactory corrosion resistance at the boiling point of these reagents.

The passivation of the material which was manifest in all tests is a characteristic and highly advantageous property. With repeated treatments the dissolution of the material decreased, so that after 48 hours the material reached towards all reagents the resistance level of Class 1 of international corrosion specifications, with the exception of nitric acid, towards which this state was attained only after 60 hours.

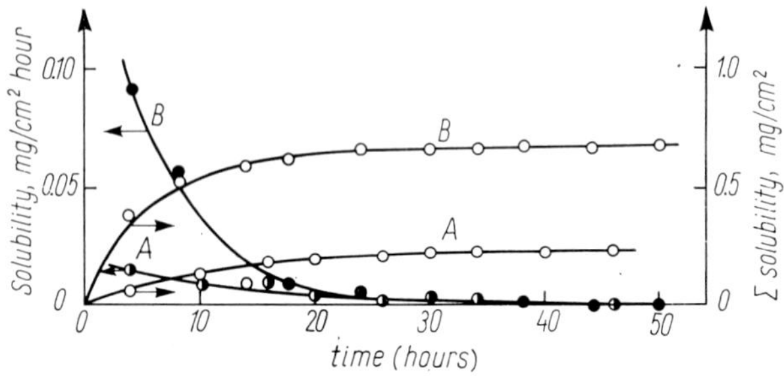

FIG. 126. Specific and gross dissolution of vitroceramic materials type KM2 in 10% and 20% sulphuric acid at boiling temperature vs. the period of treatment

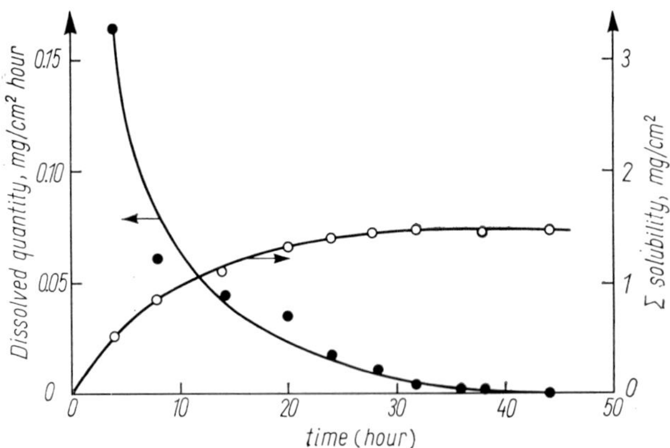

FIG. 127. Specific and gross dissolution of vitroceramic materials type KM2 in 10% sodium hydroxide solution at boiling temperature vs. the period of treatment

In certain cases the acid resistance of vitroceramics containing lithium and micro-eutectic additive and nucleated with titanium dioxide is somewhat reduced because of crystallization. This reduction is not significant, e.g. in boiling hydrochloric acid the dissolution of a vitroceramic containing 4% of Li_2O and Na_2O cannot be measured in the vitreous state even after 5 hours, while after crystallization it is $0.6 \, g/m^2 \times 24 \, h$. The micro-eutectic additive consisting of eight different oxides (ZnO, ZrO$_2$, K$_2$O, MnO, Fe$_2$O$_3$, CaO, BaO, V$_2$O$_5$) plays an important role in the composition and has a great influence on chemical resistance, that is on the decrease in acid resistance. Vitroceramics without the additive had a lower acid resistance of 28 g/m$^2 \times$ 24 h [171], that is the quantity of dissolved material was considerably higher.

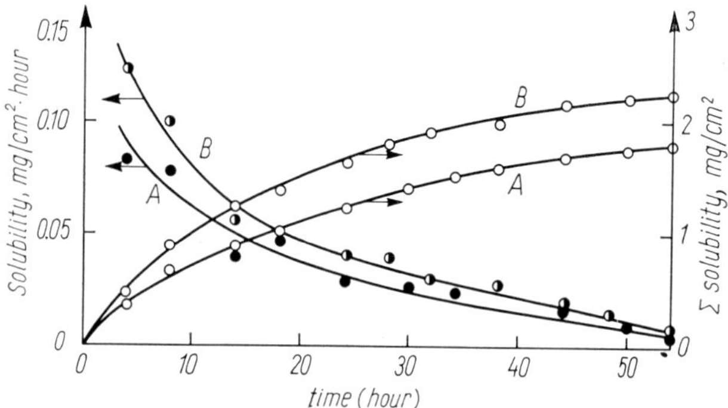

FIG. 128. Specific and gross dissolution of a vitroceramic material of type KM2 vs. the period of treatment in 10% and 20% nitric acid, resp. at the boiling temperature of the acids

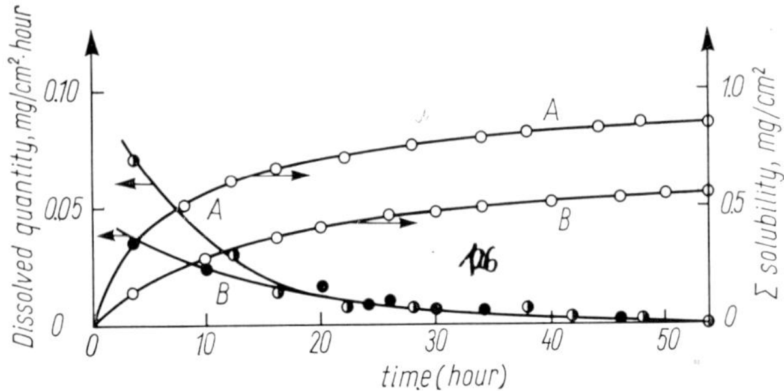

FIG. 129. Specific and gross dissolution of a vitroceramic material of type KM 2 vs. the period of treatment in 10% and 20% hydrochloric acid resp. at the boiling temperature of the acids

The application possibilities of molten silicates are far from being exploited. As with all new structural materials their utilization begins in two ways. One is the replacement of more expensive or scare materials, the other is in new constructions which could not be achieved with the earlier known types. From the aspect of their production vitroceramics are part of glass technology, while the products have the nature of ceramics. Crystallization of the glass opens new possibilities, that is ways of new applications which are based at least partly on the fact that compared to ceramics the low flexural and tensile strength as an application impeding factor is less dominant in the case of vitroceramics.

Applicability is influenced by formability and consequently the applicability of molten silicates also varies. Molten rocks can be molded only by casting, while for other vitroceramics the molding possibilities are far more manifold.

Applications are determined mainly by the three following properties of the material:

1. mechanical properties, first of all abrasion resistance
2. chemical resistance
3. dielectric properties.

Because of its high abrasion resistance molten basalt is extensively used for linings and coatings to replace iron and lead. Large quantities are utilized as pipe lines, gutter and tank linings. The molten basalt pipes are joined either by various binders, or embedded in cement paste incorporated in flanged iron pipes and the lines joined by means of the flanges. In ore processing molten basalt is used for mud channels. Experience concerning the lifetime of molten basalts shows that while similar metal pipe lines have a lifetime of 6–8 weeks, the molten basalt pipes will last up to one year. In mining and coal processing molten basalt is used in the chutes and to transport coal dust. These materials are also applied as linings of coke, coal, ore and gravel storage tanks, where the application of molten silicates quite unambiguously raises the useful life of the tanks. Because of their high abrasion resistance and toughness molten silicates may be used as floor coverings where the floor is exposed to high abrasive wear and acid-alkaline corrosion. Compared to cement flooring the lifetime of molten silicates is 6–10 times longer [2, 3, 4, 19, 158, 169]. The objects shown in the chapter on the molding of vitroceramics on feldspar–diopside base are but examples of the wider applicability of these materials.

A rather surprising new application of the pyroceramic type materials is the conic nose of missiles. This part of the rocket is exposed to very great temperature fluctuations associated with dynamic air pressure, erosion and vibration. The conic stream-lined missile sheath has the advantage of protecting the control equipment against influences from the outside enabling the direction of the missile also by its constant dielectric properties. It has the further advantage of low specific weight and possibilities of mass production. Pyroceramics require special processing by centrifugal casting which, however, provides for the required tolerance of the dimensions. By this molding technology, products are obtained after heat treatment which may be finished by polishing. High dimensional accuracy is one of the crucial points in production and this requires polishing which is complicated by the excessive hardness of vitroceramics. For this reason products designed for applications of this type requiring high dimensional accuracy are polished when in the vitreous state, so that adjustment to the final dimensions requires only slight machining in the crystalline state.

As every other structural material vitroceramics also involve a number of assembly problems on whose solution the applicability of vitroceramics depends to no small degree.

Pipes produced by glass technological methods have 3–10 cm diameters and are 1–1.5 m long. These vitroceramic pipes are highly recommended for heat exchangers operating at high temperatures partly because of their favorable corrosion resistance and partly because of their heat conductivity and heat transfer capacity which are better than those of glass. The excellent heat resistance of vitroceramics – some of which may be used up to temperatures of 1300–1350 °C – offers possibilities of constructing high temperature heat exchange equipment. Electrically heated spirals of these pipes may be utilized to heat aggressive materials.

Reactors for the chemical industry may also be built of vitroceramics where especially the types with low thermal expansion coefficients are favorable owing to the corrosion resistance, heat shock resistance and mechanical strength of the material. Because of their high corrosion resistance vitroceramics may be used in petroleum refineries where high chemical resistance at high temperatures is a special advantage. Pumps and valves for the transport of acids may be prepared of vitroceramics, just as compressor and gas turbine blades where the high abrasion and corrosion resistance of the material are excellent recommendations. In laboratory equipment the vitroceramics have the combined advantage of high corrosion, heat shock and abrasion resistance.

Owing to the dielectric properties of vitroceramics reactor heating may be solved by coating with semiconductor glass. High abrasion resistance, combined with low thermal expansion coefficient and easy moldability recommend vitroceramics to bearing production. Abrasion resistant balls and bearing blocks are being made of vitroceramics. Because of their low heat expansion coefficient vitroceramic bearings may operate within certain temperature and load limits even without lubrication. Interesting perspectives are opened to the utilization of these new structural materials in the design of turbines and pumps where with their help more favorable constructions may be achieved.

Based on their good mechanical strength, hardness, corrosion resistance and low expansion coefficient experiments are now in progress to utilize these materials in Diesel and other internal combustion engines, where the difference between the thermal expansion coefficients of the piston and cylinder wall materials and their corrosion resistance towards the effluent combustion products offer desirable advantages. The low thermal expansion coefficient holds even the promise of motors of special types with vitroceramic cylinders and pistons.

In addition to the above, experiments are in progress in many other fields to utilize the favorable properties of vitroceramics. To mention but a few: thread guides in the textile industry, lining of mortars and mills, construction of bodies subjected to friction and grinding are only some examples where vitroceramics may bring about considerable achievements. In the form of grains it may be used as an abrasive and polishing material, to bind corundum or silicon carbide grains and to form polishing tools.

The material is very hard, has a high impact strength and has a favorable influence on the useful life of polishing and buffing devices. Vitroceramics may be used in the molding of plastics, in the production of man-made fibres (as drawing dies).

Because of the high softening point of the material it may be used as the structural material of devices where glass – due to its lower softening point – can no longer be used. Specially interesting applications of this type are open to transparent microcrystalline vitroceramics. Dinner services and other household goods of vitroceramics are being produced by several firms.

Flame resistant household goods are made mainly of low expansion type materials, where beside high heat shock resistance high impact strength, as one of the most important mechanical properties from the aspect of this particular application, opens wide perspectives to this new material. Vitroceramics are being produced also with glazed finish. The expansion coefficient and baking temperature of the glazing determine the quality of the material. The baking temperature of the glazing must not exceed the liquidus temperature of the material, or more expediently the lower limit of the crystallization range. There are yet many new solutions unexploited in this respect.

One way to solve the above problem is to direct crystallization in such a way that the surface of the body shall remain in a few micron thickness amorphous. In the other solution the glazing is a vitroceramic which is transparent after crystallization, so that the glazing may have similar properties as the vitroceramic forming the body. Application of transparent or opaque enamels in the ceramic industry opens some unexpected perspectives. It is a well known fact that the strength of electric porcelain insulators is significantly improved by the presently applied glazings, thus even better results may be expected from coatings whose flexural and tensile strength is considerably higher.

Vitroceramic metal coatings are some of the most recent developments. Essentially the metal to be coated is first covered by a primer enamel which acts as intermediary between the vitroceramic applied on the top of the enamel primer and the metal. Special vitroceramics had to be developed for this purpose, since the vitroceramic has to be applied to the base enamel as a new layer in the vitreous state to provide for an adequate bond between the layers which should be coherent and even. The layer on the base enamel is then crystallized by heat treatment. The advantages of the microcrystalline structure appear here too in the particularly favorable mechanical properties. This favorable effect is manifest from the practical aspect first of all in a several orders higher impact strength which ensures a promising future to vitroceramic coatings also in the enamel industry.

Dental prosthesis made of vitroceramics is an entirely different field of application, though in principle it has many common features with the foregoing. The elaboration of diverse solutions is now in progress and it may be expected that similarly to enameling here too the favorable mechanical properties and corrosion resistance of vitroceramics will soon be apparent with added pleasing appearance and ease of tinting which are very important factors in dentistry.

The basic condition for the application of vitroceramics as enamels is the adjustment of their heat expansion coefficient and a possible reduction of the liquidus temperature.

Their corrosion resistance makes vitroceramics highly suitable for water supply pipes made usually of iron when they have a very short lifetime, or of non-ferrous metals which are difficult to procure because of their shortage on the market.

The production of large plates based on blast furnace slag was solved in the Soviet Union through continuous rolling in black and with addition ZnO in white color. This product has a great perspectivity both in building and chemical industry.

The application possibilities of vitroceramics are far from being exhausted; we are dealing here with a new material which already shows wide and promising application perspectives.

9

References

1. FOUQUÉ and MICHEL-LÉVY, in LŐCSEI, B.: *The silicate chemical basis of the preparation of crystallized synthetic stone* (Thesis, in Hungarian) Budapest, 1956
2. GOLLOW, in GINZBURG, A. C.: *Nemetallicheskie iskopaemye* **16** 7 (1943)
3. FOUGUÉ et MICHEL-LÉVY, in PORTEVIN, A.: *Le basalte fondu*. Mém. Compt. rend. Soc. Ing. Civ. de France. 1928, 266—300
4. McMILLAN, P. W.: *Glass-Ceramics*. Academic Press, London, New York, 1964
5. RÉAUMUR, M.: *Mémoires de l'Académie des Sciences*. 1739, pp. 370—388
6. KNAPP, O.: *Crystallization of Silicate Glasses*. Akadémiai Kiadó Budapest, 1964 (in Hungarian)
7. RISSE, K.: *Archiv f. d. Eisenhüttenwesen* **3** 437 (1930)
8. KEIL, F.: *Hochofenschlacke*. Düsseldorf, 1949
9. ORMONT, B. F.: *Acta Physicochimica U.R.S.S.* **20** 503 (1945)
10. ORMONT, B. F.: *Acta Physicochimica U.R.S.S.* **20** 737 (1945)
11. ORMONT, N. N.: *Vestnik Moskovskogo Universiteta* **5** 117 (1950)
12. FENNER, C. N.: *Amer. J. Sci.* **18** 219 (1929)
13. JUGOVITS, L.: *Magyar Technika* **1** 35 (1949)
14. BARTH, T. F. W.: *Amer. J. Sci.* **31** 321 (1936)
15. HINZ, W.: *Silikate. Einführung in Theorie und Praxis*. Veb. Verlag für Bauwesen, Berlin 1963
16. EITEL, W.: *The Physical Chemistry of the Silicates*. The University of Illinois. Chicago Press, Chicago 1954
17. PELIKAN, A.: *Molten rocks*. Nakl. Prace, Praha, 1955 (in Czech)
18. VOLDÁN, J.: *Sklar a keramik*. **14** 220 (1964)
19. NEBRENSKIJ, J. and VOLDÁN, J.: New methods for the crystallization of glass. *Informationi prochled SVUS* (1961) (in Czech)
20. LŐCSEI, B., POLINSZKY, K., SCHLIESS, J. and SOLTÉSZ, E.: Hungarian Patent No. 243, 012/1952
21. LŐCSEI, B., SOLTÉSZ, E., FODOR, E. and VÁRTAY, E.: Hungarian Patent No. 143.041
22. LŐCSEI, B. P.: Hungarian Patent No. 146.137
23. LŐCSEI, B. P.: DRP 1.085.804
24. WAGNER, H.: DRP 863.176
25. STOOKEY, S. D.: DRP 962.110
26. STOOKEY, S. D.: DAS 1.045.056
27. STOOKEY, S. D.: *Ceramic Fabrication Process* **21** 189 (1958)
28. STOOKEY, S. D.: *Engng. Chem.* **51** 805 (1959)
29. STOOKEY, S. D.: USA Patent 2.515.939
30. BECKER, H.: DRP 410.351
31. BECKER, H.: DRP 430.387
32. LUNGU, S. N. and POPESCU-HAS, D.: *Studii si cercetari de chemie* **7** 225 (1955)
33. LUNGU, S. N. and POPESCU-HAS, D.: *Épitőanyag* **10** 86 (1958)
34. MAURER, R. D.: *J. Appl. Phys.* **29** 1 (1958)
35. CLAYPOOLE, S. S.: Brit. Patent, No. 822.272 (1959)
36. McMILLAN, P. W. and HODGSON, B. P.: Brit. Patent, No. 31718 (1959)
37. McMILLAN, P. W. and HODGSON, B. P.: Brit. Patent, No. 944.571 (1963)
38. McMILLAN, P. W. and PARTRIDGE, G.: Brit. Patent, No. 8738 (1959)
39. McMILLAN, P. W. and PARTRIDGE, G.: Brit. Patent. No. 924.996 (1963)
40. KITAIGORODSKY, I. I. and BONDARYEV, K. T.: *Steklo i Keramika* **20** 1 (1963)
41. RINDONE, G. F.: *Amer. Cer. Soc.* **41** 41 (1958)

42. ALBRECHT, F.: *Beispiele angewandter Forschung. Neuartige Hartstoffe aus Glas.* München, 1955 pp. 19—22
43. BÁRTA, R.: *Silikáty* **2** 296 (1958)
44. HINZ, W.: *Silikattechnik* **10** 596 (1959)
45. KOPECZKY, L. and VOLDÁN, J.: *Crystallization of molten rocks.* Prague, 1959
46. LILLIE, H. R.: *Glass Technology* **1** 115 (1960)
47. LIN, F. C.: *Glass Ind.* **40** 717 (1959)
48. LUNGU, S. N. and POPESCU-HAS, D.: *Industria Usoara* **2** 64 (1958)
49. LUNGU, S. N. and POPESCU-HAS, D.: *Silicates Industries* **22** 391 (1958)
50. MORIYA, T., SAKAIONI, T., SAINO, H. and ENDO, M.: *J. Cer. Ass. Japan* **68** 44 (1960)
51. MUNIER, J. H.: *Proc. Engng.* **29** 87 (1958)
52. RINDONE, F. E.: *J. Amer. Ceram. Soc.* **45** 7 (1962)
53. ROSENBERG, W.: *Mach. Design.* **31** 29 (1959)
54. SIMPSON, H. E.: *Glass Industry* **40** 13, 17 (1959)
55. SELYUBSKY, V. I. and VAISFELD, N. N.: *Steklo i keramika* **17** 23 (1960)
56. TAMURA, Y.: *J. Cer. Assoc. Japan* **66** 325 (1958)
57. TASITO, H. and WADA, S.: *J. Cer. Assoc. Japan* **66** 390 (1958)
58. TRANZEN, C.: DRP 569.310
59. TURNBULL, D. and VONNEGUT, B.: *Ind. Engng. Chem.* **44** 1292 (1952)
60. BECKER, B.: DRP 630.898
61. WAGNER, H.: DRP 910.038
62. WAGNER, H.: DRP 927.978
63. WAGNER, H.: DRP 952.514
64. MAURER, R. D.: *J. Appl. Phys.* **29** 1 (1958)
65. VOLDÁN, J.: Czechoslovak. Patent 92.750
66. WEYL, W.: *Sprechsaal* **93** 128 (1960)
67. WIHSMANN, F.: *Thesis.* Bergakademie Freiberg, 1959
68. ZHUKOVSKY, E. V. and PORTUGALOV, D. I.: *Steklo i keramika* **15** 41 (1958)
69. HINZ, W. and KNUTH, P. O.: *Silikattechnik* **11** 605 (1960)
70. VOLDÁN, J.: *Silikattechnik* **7** 48 (1956)
71. BROUKAL, J.: *Silikattechnik* **13** 48 (1962)
72. KITAIGORODSKY, I. I.: *Zhurn. Vses. Khim. Obshsh. im. D. I. Mendeleeva* **8** 192 (1963)
73. ZIEMBA, B. and CHLOPICKA, E.: *Steklo i Keramika* **12** 69 (1965)
74. VOGEL, W. and GERTH, K.: *Zeitschr. Chemie* **2** 261 (1962)
75. MAURER, R. D.: *J. Appl. Phys.* **33** 2132 (1962)
76. TAMMANN, G.: *Kristallisieren und Schmelzen.* J. A. Barth, Leipzig, 1903
77. TAMMANN, G.: *Aggregatzustände.* L. Voss Verlagsh. Leipzig, 1923
78. TAMMANN, G.: *Der Glaszustand.* L. Voss Verlagsh. Leipzig.1933
79. LŐCSEI, B.: *Experiments to produce glass and synthetic stone profiles.* NEVIKI, Report No. 79. (in Hungarian) 1952
80. GROFCSIK, J.: Preparation of crystallized synthetic stone (*Five Years of NEVIKI*) Veszprém, 1954 (in Hungarian)
81. POLINSZKY, K. and LŐCSEI, B.: *Magyar Technika* **6** 216 (1954)
82. PREBUS, A. F. and MICHENER, J. W.: *Physic. Rev.* **87** 201 (1952)
83. SELYUBSKY, V. I.: *Steklo i Keramika* **11** 19 (1954)
84. OBERLIES, F.: *Naturwiss.* **43** 224 (1956)
85. VOGEL, W. and GERTH, K.: *Glastechn. Ber.* **31** 15 (1958)
86. VOGEL, W. and GERTH, K.: *Silikattechnik* **9** 353 (1958)
87. VOGEL, W. and GERTH, K.: *Silikattechnik* **9** 495 (1958)
88. SKATULLA, W., VOGEL, W. and WESSEL, H.: *Silikattechnik* **9** 19 (1958)
89. VOGEL, W.: *Silikattechnik* **9** 323 (1958)
90. VOGEL, W.: *Silikattechnik* **10** 241 (1959)
91. HOFFMANN, L. C. and STATTON, W. O.: *Nature* **176** 561 (1955)
92. PORAY-KOSHITS, E. A.: *Glastechn. Berichte* **3** 450 (1959)
93. VOGEL, W. and GERTH, K.: *Zeitschr. Chemie* **2** 261 (1962)
94. VOGEL, W.: Symposium sur la fusion du verre, Bruxelles 1958. *Compte rendu, Union scientifique Continentale du verre.* 741
95. STOOKEY, S. D.: *Glastechn. Ber.* **32** 1 (1955)

96. VOGEL, W. and GERTH, K.: Symposium on nucleation and crystallization in glasses and melts. *Amer. Ceram. Soc.* p. 11 (1962)
97. ROY, R.: Symposium on nucleation and crystallization in glasses and melts. *Amer. Ceram. Soc.* p. 39 (1962)
98. OHLBERG, S. M., GOLOB, H. R. and STRIKLER, D. W.: Symposium on nucleation and crystallization in glasses and melts. *Amer. Ceram. Soc.* p. 55 (1962)
99. RINDONE, G. E.: Symposium on nucleation and crystallization in glasses and melts. *Amer. Ceram. Soc.* p. 63 (1962)
100. LŐCSEI, B. P.: Symposium on nucleation and crystallization in glasses and melts *Amer. Ceram. Soc.* p. 71 (1962)
101. HILLIG, W. B.: Symposium on nucleation and crystallization in glasses and melts. *Amer. Ceram. Soc.* p. 77 (1962)
102. WESTMANN, A. E. R. and KRISHNA-MURTHY, M.: Symposium on nucleation and crystallization in glasses and melts. *Amer. Ceram. Soc.* p. 91 (1962)
103. VOGEL, W.: *Zeitschr. Chemie* **3** 313 (1963)
104. VOGEL, W.: *Zeitschr. Chemie* **3** 154 (1963)
105. VOGEL, W.: *Zeitschr. Chemie* **3** 271 (1963)
106. LŐCSEI, B. P.: *Interceram.* No. **2** 133 (1966)
107. FRENKEL, J.: *Kinetic Theories of Liquids.* Clarendon Press. Oxford 1946
108. LŐCSEI, B. P.: *Épitőanyag* **14** 241 (1962)
109. LŐCSEI, B. P.: *Silikattechnik,* in press
110. VOLMER, M.: *Kinetik der Phasenbildung.* Dresden–Leipzig 1939
111. ZACHARIASEN, W. H.: *J. Amer. Chem. Soc.* **54** 3841 (1939)
112. WARREN, B. E.: *Zeitschr. Krist.* **86** 349 (1933)
113. MACKENZIE, J. D.: *Modern Aspects of the Vitreous State.* Reinhold, New York 1959
114. VOGEL, W.: *Struktur und Kristallisation der Gläser.* V. f. Grundstoffindustrie, Leipzig 1965
115. McMILLAN, P. W.: *Glass Ceramics.* Academic Press, London–New York 1964
116. JEBSEN-MARWEDEL, H.: *Glastechn. Ber.* **29** 223 (1956)
117. JEBSEN-MARWEDEL, H.: *Kolloid-Zeitschr.* **137** 118 (1954)
118. JEBSEN-MARWEDEL, H.: *Naturwiss.* **45** 260 (1958)
119. JEBSEN-MARWEDEL, H.: *Kolloid-Zeitschr.* **150** 137 (1957)
120. JEBSEN-MARWEDEL, H.: *Glastechn. Ber.* **31** 431 (1958)
121. JEBSEN-MARWEDEL, H.: *Glastechnische Fabrikationsfehler.* Springer Verl. Berlin, 1959
122. VERESS, Z.: Verbal communication; 1960
123. DIETZEL, A.: *Zeitschr. Elektrochem.* **48** 9 (1942)
124. KLOBER, W.: *Silikattechnik* **13** 5 (1962)
125. GARNER, W. E.: *Chemistry of the Solid State.* Clarendon P., London, p. 159, 1955
126. GUTZOW, J.: *Zeitschr. anorg. allg. Chem.* **302** 259 (1959)
127. HINZ, W.: *Silikattechnik* **10** 119 (1959)
128. FOUQUÉ, F. and MICHEL-LÉVY, A.: *Synthèse des Roches,* Paris, 1889
129. AVGUSTINNIK, A. I.: *Silikattechnik* **12** 275 (1961)
130. SALMANG, H.: *Physikalische und chemische Grundlagen der Keramik.* Springer Verl. 1958
131. OKOROKOV, S. D., VOLFSON, S. L. and KORNEEVA, T. F.: *Rab. Leningr. Tehn. Inst. im. Lensoveta* **56** (1960)
132. TOROPOV, N. A., VOLFSON, S. L. and SICHEV, N. M.: *Rab. Konf. Cement,* Promstroiizdat 1956
133. ZHURAVLEV, V. F., VOLFSON, S. L. and SICHEV, N. M.: *Cement SSSR* **3** (1950)
134. KUKOLEV, G. V. and YALIMOVA, M. A.: *Petrografii i prikladnoii mineralogii SSSR* **2** (1953)
135. BUDNIKOV, P. P. and BEREZHNOI, A. S.: *Solid state reactions.* Promsztroiizdat, Moscow, 1949 (in Russian)
136. BEREZHNOI, A. S.: *Zhurnal prikladnoi khimii* **13** 800 (1940)
137. HEDVALL, J. A.: *Zeitschr. phys. Chem.* **33** 125 (1926)
138. BOBROVNIK, D. P. and BUDNIKOV, P. P.: *Zhurnal prikladnoi Khimii* **21** 7 (1948)
139. POLUBOYARINOV, D. N. and VIDIK, G. A.: *Doklady AN SSSR* **88** 325 (1953)
140. KUKOLEV, G. V. and DOLGIKH, G. S.: *Ogneypory* **12** 536 (1950)
141. NAZARENKO, M. F. and PAZUMOVA, V. L.: *Ogneypory* **65** 12 (1957)

142. Lőcsei, B.: *Ber. d. Deutsch. Keram. Ges.* **42** 277 (1965)
143. Kukolev, G. V. and Scheglov, S. I.: *Zhurn. Prikl. Khimii* **29** 1502 (1956)
144. Besborodov, N. A.: *Glass Ceram. Bull. India* 54 (1960)
145. Hedwall, J. A.: *Glass Ceram. Bull. India* 29 (1960)
146. Richards, R. G. and White, J.: *Trans. Brit. Ceram. Soc.* **40** 53 (1954)
147. Laydon, G. A. and McQuerrie, W.: *J. Am. Ceram. Soc.* **42** 89 (1959)
148. Lőcsei, B.: The effect of nucleators on the crystallization of glasses. *Éakki Report*, Budapest 1966 (in Hungarian)
149. Weyl, W. A. and Marbore, E. C.: *Glass Ind.* **41** 429, 487, 549, 620 (1960)
150. Neveux, V.: *Génie civil* **87** 57 (1925)
151. Drin, L.: *Chimie et Industrie* **4** 662 (1922)
152. Daly, R. A.: *Igneous Rocks and the Depth of the Earth.* New York 1933
153. Beyersdorfer, P.: *Silikattechnik* **5** 381 (1954)
154. Voldán, J.: *Sklar a Keramik* **5** 14 (1955)
155. Voldán, J.: *Sklar a Keramik* **5** 27 (1955)
156. Lőcsei, B.: Glass melting in shaft furnaces. *SziKKI Report*, Budapest 1965 (in Hungarian)
157. Lőcsei, B.: Experiments to produce crystalline synthetic stone. *NEVIKI Report*, No. 109, Veszprém 1955 (in Hungarian)
158. Berezsnoi, A. I.: Svetochyvstvitelnie stekla i steklokristallicheskie materiali tipa "Pirokeram". *VINITI* Moscow (1960) (in Russian)
159. Korach, M.: *Épitőanyag* **10** 1 (1958)
160. Hartman, F.: *Stahl u. Eisen* **53** 509 (1933)
161. Lamorf, J.: *Stahl u. Eisen* **56** 1006 (1936)
162. Endell, K.: *Stahl u. Eisen* **40** 213, 1 (1960)
163. Lőcsei, B.: *Acta Chim. Acad. Sci. Hung.* **22** 1 (1960)
164. Evstropyev, K. S. and Toropov, N. A.: *Chemistry of silicon and physical chemistry of silicates.* (In Hungarian) Nehézipari Kiadó, Budapest 1951
165. Eitel, W.: *Die heterogenen Schmelzgleichgewichte silikatischer Mehrstoffsysteme* Springer, Leipzig 1941
166. Morey, G. W.: *The Properties of Glass.* Reinhold, New York 1952
167. Osborne, E. F.: *J. Amer. Ceram. Soc.* **33** 219 (1950)
168. Volarovich, M. F. and Leontyeyev, A. A.: *Compt. rend.* **55** 241 (1947)
169. Lőcsei, B.: *Épitőanyag* **11** 247 (1959)
170. Lőcsei, B.: *Silikattechnik* **17** 249 (1966)
171. Lőcsei, B.: Unpublished data
172. Little, J. R. and DeClerk, D. H.: *Mitt. Deutsch. Emailfachleute E. V.* **14** 93 (1966)
173. Maskovich, M. O. and Varsal, B. G.: *Neorganicheskie materiali* **3** 1668 (1967)
174. Kumar, S. and Nag, B. B.: *J. Am. Ceram. Soc.* **49** 10 (1966)
175. Lajarte, S.: *Silic. Ind.* **38** 177 (1958)
176. Kitaigorodsky, I. I., Rabinovits, E. M. and Selyubszky, V. I.: *Steklo i keramika* **1** 20 (1963)
177. McDowell, J. F.: *Ind. a Eng. Chem.* **58** 39 (1966)
178. Thach-Lan, T.: *Verre. Refr.* **56** 416 (1965)
179. Bayer, G. and Hoffmann, F.: *Glass Techn.* **7** 94 (1966)
180. Little, J. R., DeClerk, D. H.: *Mittl. V. D. Email* **14** 93 (1966)
181. Lőcsei, B. P. and Karabélyos, P.: *Épitőanyag* **20** 74 (1968)
182. Fedorovsky, A. and Tikachinsky, J. D.: *Steklo. Inst. Stekla* **15** 51 (1964)
183. Anon: *Steklo i keramika* **17** 85 (1966)
184. Emrich, B. R.: *Mater. Des. Eng.* **62** 95 (1965)
185. Schleifer, P.: *Steklo i keramika* **17** 85 (1966)
186. Kivlighn, H. D.: USA Patent 3.146.114 (1964)
187. McDowell, F.: USA Patent 3.201.266 (1965)
188. Mwkherjec, S. P. and Rogers, P. S.: *Phys. Chem. of Glass* **8** 81 (1967)
189. Scharbach, H.: *Werkst. u. Korr.* **16** 2 (1965)
190. Duko, D. A., McDowell, J. F. and Karstetter, B. R.: *J. Am. Ceram. Soc.* **50** 67 (1967)
191. Beall, G. H., Karstetter, B. R. and Rittler, H. L.: *J. Am. Ceram. Soc.* **50** 181 (1967)
192. Karstetter, B. R. and Voss, R. O.: *J. Am. Ceram. Soc.* **50** 133 (1967)

193. Brit. Patent 822.272 (1966) Corning Glass Works
194. McMILLAN, P. W. and PARTRIDGE, G.: Brit. Patent. 943.599; 944.571 (1960)
195. Brit. Patent. 869.315 (1958), Corning Glass Works
196. Brit. Patent. 924.996 (1959)
197. McMILLAN, P. W., HODGSON, B. P. and PARTRIDGE, G.: *Glass Technol.* **7** 121, 128 (1966)
198. HODGSON, B. P. and McMILLAN, P. W.: *Glass Technol.* **4** 173 (1963); **5** 142 (1964)
199. TASHIRO, M.: *Glass Ind.* **46** 366 (1966)